Palgrave Studies in Literature, Science and Medicine

Palgrave Studies in Literature, Science and Medicine is an exciting series that focuses on one of the most vibrant and interdisciplinary areas in literary studies: the intersection of literature, science and medicine. Comprised of academic monographs, essay collections, and Palgrave Pivot books, the series will emphasize a historical approach to its subjects, in conjunction with a range of other theoretical approaches. The series will cover all aspects of this rich and varied field and is open to new and emerging topics as well as established ones.

Abigail Boucher

Science, Medicine, and Aristocratic Lineage in Victorian Popular Fiction

palgrave
macmillan

Abigail Boucher
Aston University
Birmingham, UK

ISSN 2634-6435 ISSN 2634-6443 (electronic)
Palgrave Studies in Literature, Science and Medicine
ISBN 978-3-031-41140-3 ISBN 978-3-031-41141-0 (eBook)
https://doi.org/10.1007/978-3-031-41141-0

Cover illustration: Lakeview Images / Alamy Stock Photo

This Palgrave Macmillan imprint is published by the registered company Springer Nature Switzerland AG.
The registered company address is: Gewerbestrasse 11, 6330 Cham, Switzerland

Paper in this product is recyclable.

To my family

Acknowledgements

This monograph, which was partially based on my PhD thesis, is indebted to the tireless patience, advice, and encouragement of my supervisors, Professor Alice Jenkins and Dr Matthew Creasy. The kind permission of the Hunterian and the University of Glasgow's Special Collections facilitated my research on the silver fork novels, without which Chap. 2 would not have been possible.

I am grateful to Dr Laura Eastlake, Dr Douglas Small, Dr Hannah Tweed, Dr Sarah Gillies, Dr Marcello Giovanelli, Dr Chloe Harrison, Dr David Shuker, and Prof. Henry Snow, who all provided notes and resources for various chapters or proposals. I would like to express a particularly large amount of gratitude to Dr Daniel Jenkin-Smith, whose discussions, debates, shared reading material, and comments on drafts over the past six years have greatly shaped and improved this work (especially Chap. 4).

Finally, I would like to thank my family for their constant support and encouragement.

An earlier and shorter version of Chap. 3 of this work was published in *Nineteenth-Century Gender Studies*, 9:2 (Summer 2013); my thanks to the editors for permission to include an updated version of it here.

Contents

1	**Introduction**	1
	Pathologising the Aristocracy Through the Gaze	7
	Genre Fiction	12
	Works Cited	23
2	**Fashionable Illness: Consumerism, Medicine, and Class in the Silver Fork Novels**	27
	Introduction	27
	Silver Fork Novels: An Overview of Genre	29
	The Silver Fork Novels, Health, and Medicine	35
	The Cult of Sensibility	36
	Disease and Death as Plot Devices	41
	Business Models	44
	Silver Fork Novels: Three Case Studies	48
	Godolphin	49
	Cheveley	57
	Cecil	62
	Conclusion	66
	Works Cited	72
3	**'Unblessed by Offspring': Fertility and the Aristocratic Male in Reynolds's *The Mysteries of the Court of London***	77
	Introduction	77
	Medicine, Morality, and Masculinity	81

Feminisation 87
Emasculation 91
Feminisation and Emasculation in Unison 94
Conclusion 101
Works Cited 107

4 'But You Know There's a Cousin': Endogamous Marriage in Sensation Fiction 111
Introduction 111
Sensation Fiction: An Overview 116
Endogamy and Ecclesiastical Issues 119
The Science of Endogamy: Nineteenth-Century Class Contexts 123
Sensational Endogamy: A Question of Race 134
Conclusion 144
Works Cited 150

5 Physiognomy, Evolution, and the Divine in Ruritanian Romances 155
Introduction 155
Ruritania, or the Chivalric Feudal 160
The Prisoner of Zenda 164
Prince Otto 169
The Lost Prince 174
Conclusion 179
Works Cited 184

6 'Nature Works On': Class Hierarchies in the Evolutionary Feudal 187
Introduction 187
After London 191
The Time Machine 200
The Purple Cloud 206
Conclusion 217
Works Cited 223

7 Conclusion 227
Works Cited 231

Index 233

CHAPTER 1

Introduction

In 1983, the historian R.F. Foster stated that in the realm of nineteenth-century scholarship, study of the working class was 'historigraphically exhausted' but the aristocracy was 'not yet academically respectable'; the middle classes would therefore be the next major locus for critical examination.[1] Foster's prediction was accurate. Now, forty years later, academic work on the aristocracy is limited, while research on the middle and working classes remains very active. This is especially true in the field of Victorian literature, where the scholastic reticence to discuss an entire social class is, frankly, remarkable. The aristocracy in the nineteenth century was 'consistently less than 1 per cent of Britain's population', and yet literature from the time disproportionately abounds with aristocratic characters.[2] Noteworthy though this critical silence may be, perhaps it is understandable—society is increasingly fatigued by the rich and powerful, of their prominence in much of recorded history (which they often wrote or patronised themselves), of the same stories being retold while other narratives and voices and traditions get ploughed over. To study representations of the aristocracy, one runs the risk of being seen as its ardent supporter or a dull thinker treading the same well-worn paths of scholarship that the upper classes themselves often carved out.

This still, however, leaves us with a gap in knowledge: as Antony Taylor argues in his 2004 historical work, *Lords of Misrule*, 'the social history of

A. Boucher, *Science, Medicine, and Aristocratic Lineage in Victorian Popular Fiction*, Palgrave Studies in Literature, Science and Medicine, https://doi.org/10.1007/978-3-031-41141-0_1

landed society was entirely disregarded [...] The role of the great aristocratic dynasties was simply acknowledged, rather than analysed.'[3] It is precisely this gap in knowledge and this lack of *critical* engagement which helps to reinforce aristocratic hegemony, to accept its large influence at face value, to take for granted its relationships with other class groups as straightforward.[4] As the field currently stands, the aristocracy as a class is largely understudied and often fenced off from studies of other classes. This segregation to some extent mirrors the aristocratic system itself, in which, as will be discussed, exclusivity and social division play a large part. Despite its real or perceived segregation, the aristocracy does not and cannot operate independently from other classes. Even the superlative nature of the terminology connects it to other groups: the Greek etymology of 'aristocracy', 'a ruling body of the best citizens', implies the existence of other citizens.[5] The terms 'upper class' and the 'upper ten thousand' imply correlating 'lower' groups. 'Exclusives', a nineteenth-century synonym for those participating in aristocratic high society, implies someone or something being excluded. Class groups in a single society are too closely enmeshed and reliant upon the existence of each other in other to ever be viewed entirely discretely. It is in nineteenth-century popular literature particularly (which was increasingly written by and for a variety of dynamic class groups) where these links are most convincingly witnessed, in which portrayals of the aristocracy served highly complex social, political, artistic, and—as I argue in this book—even medical and scientific purposes.

This monograph traces how public understandings of major scientific and medical developments in the nineteenth century found fertile ground for discussion when projected onto aristocratic characters—especially in new forms of genre and popular fiction, which lent different voices and perspectives to these conversations. This is not to say that discussions of medicalised bodies in nineteenth-century literature were limited to the aristocracy or just to the few genres explored here; nothing could be further from the truth. Literary, social, and historical scholars have amply traced the perceived or represented pathologies in working- and middle-class groups, as well as in groups based on gender, race, nationality, sexuality, profession, age, and disability status. And medicine, science, and pathology have all featured heavily in most nineteenth-century genres and literary movements: from the feverishness of the Romantics to the body-horrors of the Gothics, to the poet Robert Browning's psychosis-induced dramatic monologues, to various (usually working-class) morbidities in realist and naturalist texts. As this monograph seeks to illustrate in the coming chapters, the abundance of this scholarly landscape only heightens by contrast what's missing—especially considering how frequently the pathologised aristocratic body is featured in primary texts themselves. This work aims not only to fill a (somewhat segregated) gap in knowledge, but also to provide an additional overlay for previous, established scholarship, in which we can see how the many *metaphorical* registers which link aristocrats to scientific and medical concerns begin to blend or be reworked for other demographics.

This book is about aristocratic lineage and pathology, how they were understood in both a clinical and a metaphorical sense by a general population, and how literary representations of that visually prominent lineage could be used as a safe 'sandboxing' exercise for different classes to test out or grapple with new scientific and medical concepts. Lineage (what would come to be known more scientifically in this period as 'heredity') is a hallmark of the aristocratic institution and conceptually links together much of the political, economic, socio-behavioural, and scientific thought about the purpose, origin, and future of the aristocracy—and, by extension, of all other classes. Much of the fiction examined in this monograph looks backwards in order to look forwards. By triangulating aristocratic lineage, medicine and science, and popular literature, nineteenth-century authors and readers could stress-test new and often anxiety-inducing concepts, theorise about how their world has come to take the shape it has, and speculate on what this meant for their own bodies and their own future.

Before proceeding, I should note that I use the term 'aristocracy' loosely and synonymously with 'nobility' and 'upper class'. What is signified by the use of these three largely interchangeable terms is a socially elite, exclusive (though not necessarily homogenous) group comprising royalty, the hereditary peerage, and minor nobility (i.e. the baronetage). For the sake of clarity and scope, I mostly focus on texts featuring *titled* aristocracy, although some texts necessitate inclusion of untitled gentry if they are similarly pathologised based on their class or lineage. Dominic Lieven, following in the footsteps of David Cannadine's *The Decline and Fall of the British Aristocracy* (1990), writes in his historical study of aristocracy:

> [T]he peerage was only one section of the traditional upper class. There was also the baronetage and the broader untitled landowning gentry, all of which would have been defined as noble [...] Aristocracy and gentry were

> part of the same ruling class [...] To write a history purely of the peerage would therefore be to omit a key element in the story of how England's upper class confronted their rapidly changing society.[6]

The porousness of the gentry—often a halfway point between the middle classes and the lower aristocracy—is not a problem when factoring in Pierre Bourdieu's concept of *habitus*—the economic, social, cultural, educational, and even physical capital, that an individual possesses, which are used by society at large to assess one's background.[7] While much has been written on the ambiguous positioning of the Victorian middle classes, the evolving role of the 'gentleman', the permeability of wealth, and the vagaries of class perception, nineteenth-century literature is often far more conclusive with its Bourdieusian shorthand of an aristocrat.[8] More than a title, more than participation in high society, it is through a high visible status that aristocrats are often defined in literature: the reader, as well as other characters, scrutinise an aristocrat's physical form and through that scrutiny do they gain legitimacy.

It is because of their high visibility (both in terms of social stature and through constant inspection) and because of their perceived status as public property that aristocratic bodies could be used by the general public as barometers of any medical and scientific concern. Much as with any modern-day celebrity, aristocrats (especially beginning in the nineteenth century with the rise of mass media) were documented far more than the ordinary person, but with a special emphasis on their lineage and their physical, mental, and moral suitability to rule. While the development of formal concepts of celebrity culture is often attributed to Max Weber at the turn of the century, these concepts were well in effect (if not discussed in quite the same terms) at least a century beforehand and were beginning to be understood and toyed with in the literature of the period. To take one example from silver fork fiction, the narrator of Lady Theresa Lewis's 1834 novel *Dacre* decrees, 'It is the fate of distinction [i.e. of upper-class society] to be most often misjudged, because obscurity is not *judged* at all.'[9] Lewis, who was an aristocrat by birth, the wife of the famous novelist T.H. Lister, and a prominent author herself, voices this sentiment through a third-person narrator and may be speaking from her own experience in the public eye.[10] In 1842, an article in *The Foreign Quarterly Review* noted that 'the prestige of high-sounding names tends to give [...] an equal degree of conspicuousness and notoriety. The public gaze is bent on [them] with all the eagerness of *expectation*. High birth had already raised these very different personages to a lofty stage with multitudes crowding round as spectators.'[11]

Aristocrats, as literary characters, tend to be highly visual and physical figures: they are often recognised through their own fame, or through a family resemblance, or identified by complete strangers as members of the aristocracy through vaguely defined 'noble countenances' and stamps of 'good breeding'. The public 'reading' of an aristocrat is not solely a feature of modernity, nor one only seen in literature—Ernst H Kantorowicz's seminal *The King's Two Bodies* (1957) extensively documents the long history of the conceptual separation (or lack thereof) of a monarch's personal body from the body politic. A monarch was 'married' to their realm, a '*gemina persona*, human by nature and divine by grace', and with a 'oneness' of lineage that made a predecessor and a successor a single entity under the eternal aegis of the Crown.[12] By reading the king's body, so could his subjects come to understand their own material reality: he is the land, the economy, the culture. German sociologist Norbert Elias, in his *The Civilizing Process* (1939) and *The Court Society* (1969), examines the role of the aristocrat, citing the suitable representation of one's body under the gaze of others as one of the most significant elements for sociopolitical survival.[13] Scottish anthropologist James Frazer provides another major theoretical framework for understanding this relationship, for both nineteenth-century and modern readers, in his *fin de siècle* and Edwardian study *The Golden Bough* (1890–1915) when he describes a common primal understanding of a leader: 'His [a king's] person is considered [...] as the dynamical centre of the universe, from which lines of force radiate to all quarters of the heaven [*sic*]; so that any motion of his—the turning of his head, the lifting of his hand—instantaneously affects and may seriously disturb some part of nature.'[14] If it is believed, as Frazer argues that it was in antiquity and still is in certain cultures, that a leader's body is divinely connected to the land and the fate of the people he or she rules, it follows that that body will be monitored for public reassurance and for signs of change. A ruler's physicality becomes coded in the eyes of the people, and the coding becomes engrained in a collective cultural consciousness over time.

Given the aristocracy's significant overlaps with royalty, this process of reading bodily duality naturally extended to them as well. Aristocratic bodies were viewed, read, consumed, and (mis)interpreted by their subjects—as they had been for centuries—much like an augur searching for clues about what one ought to do and what the future had in store. But where once the 'king's two bodies' had a very real material effect on the health and safety of his subjects (history books are full of subjects suffering from the political and economic fallout of disabled, ill, or dead rulers and ad hoc dynastic lines), in the nineteenth century, these readings took on a new register. Sir Edward Bulwer-Lytton wrote in 1833,

> We live in an age of visible transition—an age of disquietude and doubt—of the removal of time-worn landmarks, and the breaking up of the hereditary elements of society—old opinions, feelings—ancestral customs and institutions are crumbling away, and both the spiritual and temporal worlds are darkened by the shadows of change.[15]

Bulwer-Lytton was not necessarily wrong, but his fears were drastically oversimplified: the aristocracy remained a stable *enough* cultural institution over the course of the nineteenth century to be used continually by authors and other commentators as a palimpsestic textual object, both retaining much collective meaning and many of the customs the aristocratic institution had gathered over the centuries, while also reworking that meaning and those customs to maintain a sense of hegemony (if not the reality of hegemony) in a shifting landscape. The aristocracy remained fascinating enough to readers and writers to be included as a prominent literary subject across most genres, especially as it concerned understanding the health and future of the general population.

Working from Eric Hobsbawm's concept of the 'invented tradition', a largely nineteenth-century practice in which 'ritual and symbolic complexes' around the upper classes are invented or formalised through repetition 'to establish continuity with a suitable historic past', I argue that many of the invented traditions expressed in the literature of the age have a long-standing basis in historical class traditions, albeit adapted (sometimes beyond recognition) for a modern context. As the nineteenth century wore on, the 'king's two bodies' became compounded with new types of media (which found new, ever-present ways to update a reader on a subject), the rise of celebrity culture (which quickly included the aristocracy under its remit and absorbed or transformed aristocratic traditions of surveillance), an emergent (and alternatingly insurgent and toadying) middle class, a series of political revolutions, succession crises, and sociopolitical reforms which reduced aristocratic hegemony, and the development of several new scientific disciplines and medical concerns. It is the last of these with which this monograph is chiefly concerned: how the 'king's body' was repackaged from a more mystical, metaphorical understanding of the body politic, into a space onto which contemporary medical and scientific debates could be projected and worried over—worried over sometimes to the point of these bodies becoming pathologised.

Pathologising the Aristocracy Through the Gaze

In her work on understanding medical 'truth' and subjectivity through literary means, Christine Marks writes that, generally, in Western medicine,

> the body is constructed as chartable and controllable territory. The absence of biological proof leaves the person suffering from an invisible illness in an epistemological gap. Somatic symptom disorders violate Western society's desire for the transparent body—a desire that has grown more pronounced with medical progress.[16]

That medical controllability and accountability, especially as it concerns an aristocratic body, comes primarily through the public gaze. In 1963 *The Birth of the Clinic*, Michel Foucault traces the rising importance that the gaze held in nineteenth-century medicine; he argues that it was the gaze of the medical practitioner that gave authority to a diagnosis, beyond patient testimony or more informal interpretation of symptoms. Instead of relying on the gaze of merely 'any observer', we need the gaze 'of a doctor supported and justified by an institution [...] of a doctor endowed with the power of decision and intervention'.[17] Without denying the role and influence of the doctor and the increasing professionalisation, standardisation, and authority of the medical field through the nineteenth century, the increasing medical and scientific literacy of general populations enabled literature of the period to do something far more complex with a publicly co-opted medicalised gaze. This literacy—which Gillian Beer attributes to a clarity in scientific writing, enabling 'the reader to comprehend scientific problems well beyond his or her intellectual reach'—along with the visual prominence and documentation of aristocratic bodies and lineage, allowed the average person to map onto those bodies their own concerns, ambitions, fascinations, and (sometimes flawed) understandings of contemporary big issue debates and to watch those debates play out in real time as they were interpreted upon those bodies.[18] David Hillman and Ulrika Maude write:

> Authorities (medical and socio-economic and political) have powerfully vested interests in *constructing* bodies in particular ways; literature, throughout the ages, works to remind us of this fact and thereby to *deconstruct* these myths, often by reinstating the delirium and the scandalousness of the body. For the body is never simply a passive depository of cultural fantasy or the workings of power; it resists all reification and fixity.[19]

The long tradition of watching social and political leaders' bodies as harbingers of coming prosperity or destruction for the realm here morphed, in conjunction with large class-based paradigm shifts: an aristocrat was no longer a synecdoche of the realm itself but of its people, or even of humanity more generally. Aristocracy, in this way, was no longer the rule of society's 'best citizens', but the (sometimes cautionary) guidance of its most prominent.

Foucault's *The Birth of the Clinic* lightly segues here into his ideas of the Panopticon in his *Discipline and Punish* (1975)—which is not to say that aristocrats were marginalised, downtrodden people brought to heel by the penalising effects of society's gaze, nor that they self-policed in fear of that gaze potentially being cast upon them. Rather, through the rise of literacy rates, print media, and the sheer absorption of scientific and medical concerns into a huge variety of popular culture, Foucault's professionalised clinical gaze could be co-opted by the lay person and compounded with a tradition already well-versed in reading and even policing aristocratic bodies. His correlation of vision with power and visibility with a trap helps us to understand part of the shifting relationship between highly visible 'public' aristocrats and the often middle-class 'private' writers and readers who projected, examined, and policed those visible aristocratic bodies, and then in turn used those interpretations to examine and police themselves, sometimes from different vantages. As John Berger, in his *Ways of Seeing* (1972), argues, the locus of the gaze 'is an act of choice' and 'we are always looking at the relation between things and ourselves'.[20] Aristocrats, who occupy a rarefied and extreme position in society, are especially useful because their position can help articulate bodily norms and ideals, to separate who we are from who we are not, what we can expect from what does not apply to us, what we want to become from what we fear we will become. Leo Braudy, in his *The Frenzy of Renown* (1986), synthesises both Foucault's and Berger's arguments to include class-based and celebrity structures. Braudy writes that 'fame is always compounded of the audience's aspirations and its despair, its need to admire and to find a scapegoat' and that 'the heart of what it meant to "go public" was to be entrapped by the gaze of others, to be reduced by their definitions, and to be forced into shapes unforeseen'.[21] Many of the shapes in which aristocrats were forced were medically or scientifically essentialist, and very often pathological.

The rise and popularisation of medical and scientific professionalism recast a centuries-old religious or superstitious interpretation of a ruler's body into one now ostensibly under a scientific-medical aegis. Instead of interpreting from an aristocrat's health and demeanour the bounty of the next harvest or a pestilence or an oncoming war, these bodies instead gave a perceived degree of medical instruction on a human level—both individualistic and universal—providing a metric for norms of the body, or by creating ideals for the ambitious, or by serving as warnings. Especially in literature, aristocratic bodies operated as a visual space onto which one could project current scientific and medical issues and stress-test them: preoccupations about health and lineage and the body and new scientific developments could be worked through and taken to their most extreme, if logical conclusions.

There are several general terms I use throughout this book: 'medicine', 'science', 'pathology', 'lineage', 'heredity', among others. The problem with defining such terms is that they have their own technical or field-specific definitions which do not always align with how they are more popularly understood, and both sets of these definitions shift over time. The nineteenth century alone saw huge variations, both professionally and popularly, in what some of these terms conveyed, as will be discussed over the course of this book. Broadly speaking, when I discuss 'science', I am speaking about broad fields of empirical, testable study in the *natural* sciences, with a strong emphasis on biology. In particular, I explore how the general understanding of the natural sciences influences and is influenced by both literature and the social sciences: many issues within the natural sciences are heavily coloured by the socioeconomic conditions in which those studies are conducted, and on whom and by whom research is done. This overlap is unavoidable in any interdisciplinary discussion of class and a technical study of the body. The literature I study in this monograph traces this dynamic between natural and social sciences, but it is ultimately the art (the field most often contrasted to 'science') with which I am concerned. This is a monograph about classed bodies and science and medicine, but my major focus is on the genres, narratives, and tropes used by scientific and fiction authors alike to wrestle with or convey scientific issues. This monograph is concerned with analysing science and medicine in literature, rather than taking a Medical Humanities perspective or borrowing from its methodologies. For purposes of scope, I will make little to no reference to the formal sciences, such as logic, mathematics, and computing. My analysis does contain some reference what are now widely discredited pseudo-sciences, like physiognomy, phrenology, and eugenics, as these were embedded into other more valid scientific theories.

'Medicine' as a separate term may therefore seem slightly redundant, as it is often considered to fall under the umbrella of the natural sciences but, as, Stephanie M. Hilger discusses, 'Medicine is not a hard science per se; rather it is an interpretive approach that combines scientific and humanities-based modes of knowledge.'[22] A distinction, perhaps without much difference, is important: the separation and interplay of the two terms 'science' and 'medicine' is useful, as it implies two vantages from which Victorian authors can interpret a biological issue. The natural sciences, or biology slightly more specifically, connote (at least in popular understandings) a broader-spectrum analysis of living organisms, whereas 'medicine' is likewise popularly understood to have, by necessity, a more personal, individual, human component. These understandings of these two huge fields are, of course, hugely reductive and inaccurate, but when the *conceptual* extremes of the two vantages meet in the middle—not at biology's supposedly broad focus on all living matter, nor at medicine's supposedly granular treatment of an individual patient—do we get the both fields attempting to rationalise and classify certain *groups*, be they a taxonomical domain, through to a specific species, or down to much narrower demographics. As will be seen, the classed body was a fraught issue which could not always be codified or explained, despite its very real interpretable effects, and nineteenth-century authors and genres that were concerned with the natural sciences often need multiple vantages to explore it.

A similar conceptual division lies between terms like 'pathology' and 'illness'. Christine Marks writes in her 'Reading the *DSM-5* Through Literature: The Value of Subjective Knowing', that 'the term *illness* itself encompasses the personal elements excluded by other terms such as *disease* or *pathology*, which delineate clinical assortments of symptoms detached from subjective experience'.[23] Jan G. van den Tweel and Clive R Taylor argue that the history of pathology as a discipline has existed in tandem with the history of any medical practice, as long as 'those gross features of disease [...] were directly visible'; as has been illustrated in this introduction, few sections of society were more 'directly visible', and therefore more subject to pathologizing, than the aristocracy.[24] Although individual illness certainly plays a role in this monograph, there is more of a necessary emphasis on pathology, of an individual illness being represented as emblematic in microcosm of a far wider-spread issue. The literature I survey is far more often concerned with a class-based study of disease, disorder, biological threat, essentialism—in other words, of various *pathologies* within or without the aristocracy, their origins, and mechanisms, of which an individual or small-group illness may be taken as a case study. The term 'pathology' has also come to be known more colloquially as the progression or path of a specific disease, which has resonance in this monograph: nineteenth-century authors were often concerned with tracing the progression of classed-based scientific and medical issues. Pathology, a word dating from the sixteenth century to indicate the general study of disease, took on additional registers in the nineteenth century to include 'abnormality or malfunction in the moral, social, linguistic, or other sphere', which, according to nineteenth-century authors, cannot be untangled from its involvement with class and health.[25]

Much as today, there were innumerable scientific and medical issues percolating in the public consciousness or expressed in artwork in the nineteenth century, and public engagement or understanding of these issues was and continues to be messy. When looking at something as large as a century's-worth of concerns about the body and lineage, projected onto an institution as long-standing as the aristocracy, it can be difficult to filter down to a reasonable number of issues and shoehorn them into individual chapters. A particular question or source of anxiety, for example, does not just appear in one discrete type of media or genre or form. Nor does this question or source of anxiety always spring up fully formed, with an easily defined epicentre and a clear date of resolution. Nor can it be separated out from the dozens or hundreds of other discussions happening in the public forum. One can only enter into analysis of such discussions knowing they are a single thread in a much larger web.

There were many topics I could have chosen for this monograph: the changing role of the doctor from yeoman to professional to gentleman; the social and moral panics around anatomy, autopsy, and vivisection; the ruling forces in magnetism and mesmerism; ideas of the self in the study of emotions, cognition, and psychology; how class interacts with medical ethics, or even specific practices like (what we now call) bibliotherapy. For purposes of scope and continuity, I chose nineteenth-century medical and scientific issues which could segue easily into each other and call back to each other, despite being ostensibly separate topics. I chose issues which were heavily in debt to longer-standing conceptions of lineage and heredity, which are entangled with the bodily essentialism (or lack thereof) of 'noble blood. In particular, these had to be issues which were best explored in nineteenth-century literature by stress-testing their metaphorical or conceptual limits on aristocratic characters or institutions *but* had real bearing on the material lives of readers from all demographics. The issues I chose needed to be large-scale enough that they were in the public consciousness and written about by a variety of authors. And finally, and as will be discussed in full in the below section, I also chose medical and scientific issues which were framed as being the most in dialogue with the forms and genres in which they often cropped up; while no single issue was the sole provenance of one genre, nor vice versa, there were some scientific and medical debates, the content and context of which lent themselves nicely to (or even helped in the development of) the genres in which they often appeared.

Genre Fiction

Antony Taylor argues, '[m]ost histories have followed David Cannadine's view that aristocracy is best understood from the top down. Books like his *The Decline and Fall of the British Aristocracy* simply take on trust the views that aristocrats articulated about themselves.'[26] I argue, however, that if aristocrats have historically had the hegemonic influence and education to promote particular views of their own social class, then this historical and literary hegemony had deeply shifted by the beginning of the nineteenth century as the middle classes rose in economic, social, and political power, and literacy rates grew. With the expansion of middle- and lower-class authorship and literary demographics came middle- and lower-class representations of aristocrats. If I am tracing widespread nineteenth-century British understandings of medicine and science through the vehicle of the aristocratic body, then this monograph will require diverse voices and perspectives. I therefore need to triangulate common understandings and usages through a more representative sample than has historically been available when discussing the aristocracy.

To use a more representative sample of authors and readers means one must use 'genre' and 'popular' fiction—terms that have historically been used disparagingly by critics and other tastemakers (usually straight, white, middle- and upper-class men) as shorthand for texts for and by those who have identities different to their own: the lower- or working-classes, women, children, queer, or non-white people. In his 'Popular Fiction Studies: The Advantages of a New Field' (2010), Matthew Schneider-Mayerson laments the lack of a clear definition of popular fiction, stating that it has usually been 'defined by what it is not: "literature"'.[27] In his 1977 *Popular Literature*, Victor Neuburg wrote that

> popular literature offers us a window—and it is certainly no more than this—upon the world of ordinary men and women in the past [...] Popular literature can be defined as what the unsophisticated reader has chosen for pleasure. Such a reader may, of course, come from any class in society, although the primary appeal of popular literature has been the poor—and, by the end of the eighteenth century, also to children.[28]

Emma Liggins and Daniel Duffy, in their *Feminist Readings of Victorian Popular Texts*, rebut Neuburg's critique as patronising and inaccurate: 'This [perspective] exposes a number of common misconceptions about popular texts and their readers, namely that the latter are predominantly lower class, have no critical faculties, and choose texts that have value only as entertainment.'[29] To Liggins and Duffy's argument I add this: there is frequently a critical suspicion of pleasure in reading and this suspicion is often rooted in the idea that an enjoyable, entertaining text cannot have any intellectual merit, analytical potential, or useful and thoughtful engagement with the world. Even assuming a genre or popular fiction text *were* thoughtlessly written and mindlessly consumed—and this is a valid critique of some examples of genre fiction which were sourced and quickly generated by publishers who want to cash in on a popular formula—the *subconscious* choices and constructions of an author, as well as whatever value a 'casual' reader finds in a genre, are themselves important and intriguing patterns for scholars to analyse. The diplomat and writer Sir Richard Faber argues in his *Proper Stations: Class in Victorian Fiction* (1971) that

> there is not much to be gleaned about Victorian class attitudes from works of fantasy or poetic imagination; from mysteries (like those of Wilkie Collins); from fashionable romances (like Lytton's *Pelham* and *Godolphin*); or from stories of well-bred family life (like Charlotte Yonge's *The Heir of Redclyffe*. The same is true of historical novels.[30]

Coincidentally, these are almost the exact genres used in this monograph to prove the precise opposite. While of course some examples of genre fiction may be formulaic, rote, trite, opportunistically created, or even populated with troubling depictions, it is by virtue of their genre fiction status that we can see old tropes from new or various vantages. Their popularity is an analytical virtue, indicating they have sparked something in the zeitgeist. In this instance, that 'something' was a series of roiling medical or scientific concerns, which found ease of projection onto the figure of the aristocrat. Len Platt, lightly supporting Faber, says of aristocracy in popular fiction:

> The presence of an aristocrat or two was a basic ingredient of the recipe that produced best-sellers, and aristocracy was used in standard and often limited ways. In many novels the world of landed privilege was not much more than a glamorous playground, to which often 'dangerous' passions could be safely removed and indulged.[31]

But the aristocracy, as we have seen, is more than a 'mere' ingredient in the popular consciousness, and genre fiction is more than a 'recipe' to generate sales. That pathologised aristocrats cropped up again and again in different genres, for and by different audiences, to express different scientific and medical concerns, across the whole nineteenth century (and beyond), tells us something far greater about the strength of the aristocratic body as a literary trope, and of the strength of genre fiction to provide scholars with an interplay of vantages.

We will see more hostility—this time from contemporaneous critics—towards all of my five selected genres in turn, but first I must actually define what I consider to be 'popular' and 'genre' fiction. These are terms I will use largely interchangeably, for the sake of variety and grace, although they take on shades of their own meaning. Genre fiction is commonly understood to be anything that is not realist fiction (sometimes tellingly referred to as 'literary' fiction). In part, the antiquated literary term 'romance' is useful in defining genre fiction, as it denotes anything outside the conventions or mechanics of contemporary life. This includes, among other things, historical fiction, anything involving the supernatural or the cosmic, or speculative worlds. Over time, this definition would come to include what we now know as science fiction and fantasy. This definition does not go far enough, however, and three of my five selected genres would not be deemed 'romances'.

Genre fiction, at its core, must have some sense of identifiable narrative formula (even if that formula is later adapted or toyed with), a recognisable tone and general aesthetic style, and engage with at least a few tropes, patterns, settings, character types, and concerns which are deemed by readers and critics and emblematic of that genre. Once these tropes and patterns begin to ossify, there is frequently an element of mimicry or parody in subsequent texts; it is for this reason that genres (which are always being newly developed or reconfigured) have the reputation of being thrilling in their genesis and early examples, before treading the same ground, generating clichés, and falling out of fashion. This is not to say that the reputation is deserved: some of the genres explored in this monograph have proved to be shockingly resilient and have lasted well into the twenty-first century (albeit updated, and often with some decades-long breaks in their popularity).

Often, but not always, genre fiction is written by and for certain demographics: sometimes to much larger and diffuse demographics (a common example would be genres targeted at women—half the population, but whose tastes have historically been considered lesser), or to much more specific subcultures. 'Popular' literature largely fulfils the same function (in that it is defined against 'literary' works), but with less critical concern about granular categorisation or demographics. Popular fiction is generally deemed to comprise genre fiction, but with a focus on broad-spectrum appeal: a work that requires no specialist knowledge or above-average reading ability to enjoy, whose content and style would likely be appealing (or at least not repelling) to a majority of the population, whose title would likely be familiar to the average person. Genre and popular fiction are, in this way, useful because their contents can be considered indicative of specific social climates. Genre and popular fiction of the nineteenth century, as has already been alluded to, was rarely, if ever, critically acknowledged despite being consumed by and generating meaning to large swathes of the population.

Genres, of any kind, are difficult to define because they both require and defy borders. While many of the genres I have selected are relatively discrete, separated by their settings, tones, purposes, decades, authors, and (occasionally) audiences, they do occasionally recall or borrow from each other, segue from or into other genres, or lie in a middle ground between two or more genre groups. Genre is most aptly (at least for this book) understood as a system that Wittgenstein refers to in his *Philosophical Investigations* as a series of 'family resemblances', rather than a single, unified set of defining features.[32] A genre may contain dozens of defining features, not all of which need to be present in every example of the text: when (subjectively) *enough* matching characteristics are noticed in a text, it may potentially be classified as part of that genre. When enough of these texts form a nebula of overlapping themes, tropes, and traits, then those traits may renegotiate the genre which connects them. It is easier to conceptualise a genre as large network of overlapping Venn diagrams, each being a commonly used trope, setting, or style; one may then zoom out further and view an individual text as existing at the centre of its own Venn diagram of overlapping genres. No doubt the genre classification of some of the texts I explore can—and should—be challenged, but this is the generic morphology which forms the basis of my textual selection and should render all of them at least viable candidates for, or distant cousins to, the genres in which I have grouped them.

The reliance on multiple genres, instead of a fuller view of just one genre, is to trace the breadth and scope of literary aristocratic pathology as an almost transhistorical phenomena—or at least a phenomena which took on new registers in the nineteenth century with the rise of genre fiction. Genres themselves are unstable and often bring about the conditions of their own demise and, parasitically, the public's preoccupation with reading aristocratic forms must make the jump to a new host-genre when it has worn out the conditions of the previous. In Sianne Ngai's 2013 analysis of Laruen Berlant's *Cruel Optimism* (2011), Ngai writes

> that when the institutional supports of a fantasy or attachment start to break down, *slowly* [...] the genres that a culture uses to make sense of its present also undergo a similar diffusion, fraying, and unravelling [...] Diffusion, fraying, unraveling—slow processes of attrition or wearing-out or exhaustion—are not, for all their slowness and negativity, non-happenings; they are particular forms and not the antithesis of change.[33]

These frayings are part of the point, which each genre to some extent decrying the end of the aristocracy, but themselves dying out before our need for aristocracy-as-metaphor does.

Of all the possible nineteenth-century genres to choose—and there are several, given the famous 'rise' of the novel in this period and its immediate fracturing off into genres—I have chosen those that can cover a majority of the nineteenth century (the earliest starting in the 1820s, the latest appearing at the *fin de siècle*) and can be roughly divided into twenty- or thirty-year periods. This is to ensure maximum coverage while tracing the pathologised use of the aristocratic body: for it to crop up again and again in disparate, largely discrete, and hugely popular forms of literature illustrates the lasting resonance of the aristocratic body as a codeable symbol in the general consciousness. To give each genre and time period roughly equal weight, this meant I had to discount some genres, like Gothic fiction (and its many offshoots), Romantic fiction, drawing room novels or novels of 'manners', or detective fiction (which is a genre in its own right, but covered in part under the remit of 'sensation fiction'), as these either had an initial popularity which extended far beyond a rough twenty- or thirty-year window or experienced a popular resurgence within the nineteenth century itself. Such long timescales would render a chapter too unwieldy and would undermine the structure of the monograph.

The genres I have chosen (as will be broken down fully below) are: the 'fashionable' silver fork novels of the 1820s–1840s, the radical Chartist penny fiction of the 1830s–1850s, sensation fiction from the 1850s–1870s, and then two mutually opposed but intertwined *fin de siècle* genres, Ruritanian romances and a genre I have named the 'Evolutionary Feudal', both from the 1880s–1900s. I address these genres chronologically, not only because many of them build on each other but also for ease of contextualising developments in the perception of medicine, science, and the aristocratic body to my readers. I have stuck with genres that were developed in the nineteenth century and popular in Great Britain in this period. For purposes of scope, I largely only analyse authors from the British Isles. There is the odd and brief look at Irish or American texts, but these are infrequent examples from authors who have notably engaged with a particular genre. To examine the views of authors and readers whose entire aristocracy is imported or foreign would open up this monograph to more questions and complications than it could feasibly attempt to address.

My chapter structures vary slightly. As each genre is different in its development, author and audience demographics, portrayal of aristocratic lineage and pathology, breadth and depth of popularity, and critical coverage and reception, I have therefore allowed my chapter structures to be determined by these considerations. This flexibility enables me to not only do the most justice to the genres, but also conceptually illustrate or mirror some of the mechanisms or dynamics of these genres in my coverage of them. In my first chapter on silver fork fiction, I tightly weave together a network of texts by several authors, because the texts themselves formed a tight, highly formulaic network, which mirrors the perceived exclusivity of the genre's author- and readership; this tight network narrows in further to look at three 'case studies' from the genre in more depth. In my second chapter on the radical Chartist penny fiction, I focus on a single (but long-running) serialised narrative by one author, G.W.M. Reynolds. This is purely to keep my analytical scope manageable; the text itself is so colossal and has so many intertwined plot lines that it could equal perhaps a dozen individual novels and, while there were certainly other authors operating in the same genre, G.W.M. Reynolds was such a leader of this genre (as both a writer and a publisher) that this one work can easily stand in microcosm of the genre as a whole. My third chapter on sensation fiction slightly calls back to the first on silver fork fiction, but here I *loosely* weave together a network of texts by several authors; this loose network speaks to the genre's concern with class invasion and its reputation and lineage as a 'mongrel' genre.

My last two chapters, the first on Ruritanian romances and the second on the 'Evolutionary Feudal' necessarily take a decidedly different structure. These two chapters are heavily interlinked and form two parts of a greater discussion; they are wildly different, almost mutually exclusive genres (which are rarely, if ever, compared to each other), but they exist in a sort of Janus-faced dialogue around the issue of evolutionary biology in the aristocracy. These are the only chapters where I examine texts individually and largely sequentially from publication date, rather than as a more conceptual cluster or network. This is because, firstly, these two genres are generally smaller and have far fewer texts to analyse than something as enormous as silver fork or sensation fiction, whose entrants number in the hundreds or thousands. Secondly, the texts in both Ruritanian and Evolutionary Feudal fiction—appropriately, given their focus on evolution—all heavily build off the ones that came before them and show the genre evolving (or not) in a way that is far more conscious and linear than other genres. It would be a disservice to the genres not to follow that thread in my own structuring of the analysis. Finally, it should be noted that, while the first four genres examined in this book are all well-established and defined, the final chapter on the Evolutionary Feudal is the only one in which I define my own genre. This is because, while the greater genre of post-apocalyptic literature was still eddying around a clear identity that would only become fully actualised in the mid-twentieth century, these early examples have such a surprising level of cohesion and sophistication that it is remarkable they have only ever been identified as a sort of precursor to a genre instead of a (sub-)genre in their own right.

What follows is a quick overview of the content and argument of each chapter, now that the structure has been justified.

My first chapter explores the silver fork or 'fashionable' novels of the 1820s–1840s, ostensibly written by those either in the aristocracy or with solid connections to it. The novels exist as a sort of behavioural guide and purchasing catalogue for aspirational middle-class readers and the novels offer a great deal of advice on all subjects (some of it earnest and correct, much of it outdated, satirical, or completely wrong). One of the more intriguing threads of advice centres on the relationship between the ideal aristocratic body and ill health, and if those pathological expressions are essential to aristocratic bodies, or if they are merely learned and performed. This chapter traces the lineage of performative ill health in silver fork fiction through the older literary mode of 'Sensibility', through the genre's formal use of death and disease as plot devices, and through silver fork publishers' borrowing of, or mirroring, business practices in the early nineteenth-century healthcare industry (including false advertising, 'quack' professionals, and product tie-ins). Due to the sheer number of silver fork novels, the chapter begins with a general survey of theory and literature which triangulates the genre, aristocracy, and medical concerns around class-based ill health. The chapter ends on three case studies of notable silver fork texts—Sir Edward Bulwer-Lytton's *Godolphin* (1833), Rosina Bulwer Lytton's *Cheveley, or a Man of Honour* (1839), and Catherine Gore's *Cecil* (1841).

G.W.M. Reynolds's *The Mysteries of the Court of London*, the subject of my next chapter, examines an overt commentary on aristocratic male fertility and its relationship to biopolitics and morality. Reynolds's text dramatises the reproductive failures of aristocratic bodies and represents them as a political rallying point for his lower-class readers. Issues of aristocratic fertility, when coupled with medical literature contemporary to Reynolds's writing, reveal the innovative approach Reynolds adopted in his political argument against primogeniture. His portrayal of the male aristocratic body as reproductively doomed creates a complex space where subjective issues such as morality and gender dynamics are portrayed as biological fact.

The next chapter covers sensation fiction's concerns with aristocratic inbreeding and how endogamy or exogamy figure into the greater debates around the burgeoning fields of genetics, eugenics, criminology, psychology, and sociology. This exploration of sensation fiction illustrates how the long-standing theological, economic, and dynastic considerations of elite groups marrying within a tight kinship or social circles were reframed during the nineteenth century: these considerations now became (pseudo-) scientific ones about fertility, ableism, and the effects of race- and class-tainting. Much like with the silver fork chapter, the sheer number of sensation fiction candidates necessitate a looser 'network' structure of examination to show the genre's overall concern with, but ambivalence of, aristocratic intermarriage and the supposed safety and knowability of an endogamous group. Unlike the silver fork chapter, sensation fiction will not have any individual case study texts, but will rather give roughly equal space to as many sensation texts as possible. This enables a looser structure that can mirror the genre's own looser generic definitions and formula and can also account for the much longer timeframe that sensation fiction was technically running. While this chapter tries to focus on the years in which sensation fiction was the most popular (the 1860s and 1870s), it has strong antecedents from as far back as the late 1840s and limped on until the 1890s.

The Ruritanian and the Evolutionary Feudal genres both have aristocratic evolution and heredity at their cores, but their narratives address concepts of leadership and Darwinism from diametrically opposite positions. For this reason, these genres are examined in relation to each other in order to strengthen the perspectives and patterns found in each. Each genre uses the aristocratic body and its physical lineage as a way of addressing concerns about the impending future and understanding social origins. The first part discovers an anti-evolutionary, pro-Carlylean view of the Ruritanian aristocrat as a divinely appointed leader whose physicality and heredity are somehow outside of time and evolution. This chapter is then contrasted by the Evolutionary Feudal, whose post-apocalyptic texts place the entire aristocratic system within the confines of natural law and evolution, stating that a contemporary position of hereditary power is the result of an ancestor whose 'superiority' is rooted in the animal world: those who were the 'fittest' to survive and lead, or perhaps those whose bodies were merely in the right place at the right time. This linked approach to the aristocratic body through Darwinian critique provides the opportunity to compare two otherwise divergent genres.

Writing in his work on nineteenth-century heredity, Maurizio Meloni traces the surprising *lack* of a centralised view of lineage until the mid-eighteenth century and that, instead, 'Views of heredity were scattered among philosophical commentaries, encyclopedias, medical and moral treatises, and other sources', with similarly incohesive conclusions.[34] These disjointed views would continue well through the nineteenth century (with increasing anxiety, as Meloni argues) until the matter was largely resolved in the public consciousness in the 1890s and early 1900s. This is not to say that as a metaphor, heredity has ever been neatly packaged or settled. As the ensuing chapters will illustrate, the nearer a resolution was reached on the scientific and medical front, the more allegorical potential lineage had within literature. This is particularly true when looking at the figure of the aristocrat, whose body and heredity were subjected to the palimpsestic, cumulative readings of history and whose identities (which, by definition, looked backwards for legitimacy) were now subject to the ambiguous effects of modernity.

Notes

1. R.F. Foster, 'Tory Democracy and Political Elitism' in *Parliament and Community*, ed. by Art Cosgrove and J.I. McGuire (Belfast: Appletree Press, 1983), pp. 147–175 (151).
2. Jennifer Newby, *Women's Lives: Researching Women's Social History 1800–1939* (Barnsley: Pen & Sword, 2011), p. 109.
3. Antony Taylor, *Lords of Misrule: Hostility to Aristocracy in Late Nineteenth- and Early Twentieth-Century Britain* (Houndmills, Basingstoke: Palgrave Macmillan, 2004), p. 3.
4. There was, as Taylor's work illustrates, a very slight upsurge in academic work on the aristocracy in the 1990s and early 200s, although this was largely accomplished by Taylor himself and a few other historians, most notably: David Cannadine, *Aspects of Aristocracy: Grandeur and Decline in Modern Britain* (London: Yale University Press, 1999); Lawrence James, *Aristocrats* (2009) (London: Abacus, 2010); Stella Tillyard, *Aristocrats: Caroline, Emily, Louisa, and Sarah Lennox, 1740–1832* (London: Vintage, 1995); and, far earlier in the twentieth century, Norbert Elias, *The Civilizing Process* (1939), trans. by Edmund Jephcott (1978), ed. by Eric Dunning, John Goudsblom and Stephen Mennell (Oxford: Blackwell Publishers, 1994, repr. 2000), and *The Court Society* (1969), trans. Edmund Jephcott (Dublin: University College Dublin Press, 1983; repr. 2006). Their expansion of the field, while crucial and ground-breaking, is largely restricted to an historical approach.
5. 'aristocracy, n.', *OED Online* (Oxford University Press, June 2015) http://www.oed.com/view/Entry/10753?redirectedFrom=aristocracy#eid [accessed 7 July 2015].
6. Dominic Lieven, *The Aristocracy in Europe, 1815–1914* (Houndmills, Basingstoke: Macmillan, 1992), p. xiii.
7. Pierre Bourdieu, *Distinction: A Social Critique of the Judgement of Taste* (1979), trans. by Richard Nice (1984) (London: Routledge, 1984, repr. 1999), p. 2.

8. For comprehensive examinations of this concept, see Marie Mulvey-Roberts, 'Introduction' in *Cheveley, or The Man of Honour* by Rosina Bulwer Lytton (1839), ed. by Marie Mulvey-Roberts, *Silver Fork Novels, 1826–1841*, 6 vols, series ed. by Harried Devine Jump (London: Pickering & Chatto, 2005), V, pp. ix–xxvii (xxiv); Gwen Hyman, *Making A Man: Gentlemanly Appetites in the Nineteenth-Century British Novel* (Athens, Ohio: Ohio University Press, 2009); David Castronovo, *The English Gentleman: Images and Ideals in Literature and Society* (New York: Frederick Ungar Publishing Company, 1987); and Robin Gilmour's *The Idea of the Gentleman in the Victorian Novel* (London: George Allen & Unwin, 1981).
9. Lady Theresa Lewis, *Dacre: A Novel*, ed. by The Countess of Morley, 3 vols (London: Longman, Rees, Orme, Brown, Green & Longman, 1834), I, p. 150, emphasis mine.
10. D.A. Smith, 'Lewis, Lady (Maria) Theresa (1803–1865)', in *Oxford Dictionary of National Biography* (Oxford: Oxford University Press, 2004), Online edn., October 2006 http://www.oxforddnb.com/view/article/16595 [accessed 13 January 2015].
11. Antonia Gallenga, 'The Aristocrat of Italy', *The Foreign Quarterly Review*, 28:56 (January 1842), 362–397 (p. 362, emphasis mine).
12. Ernst H. Kantorowicz, *The King's Two Bodies: A Study in medieval political Theology* (1957) (Princeton: Princeton University Press, 2016), pp. 221–23; p. 87; p. 338.
13. Elias, *The Court Society*, p. 71; p. 116.
14. James Frazer, *The Golden Bough*, 1922 edition (London: Penguin, 1996), p. 202.
15. Sir Edward Bulwer-Lytton, *England and the English*, 5th ed. (Paris: Baudry's European Library, 1836) p. 237.
16. Christine Marks, 'Reading the *DSM-5* Through Literature: The Value of Subjective Knowing', *New Directions in Literature and Medicine Studies*, ed. Stephanie M. Hilger (London: Palgrave Macmillan, 2017), pp. 165–79 (p. 166).
17. Michel Foucault, *The Birth of the Clinic: An Archaeology of Medical Perception* (1963), trans. A. M. Sheridan (1973) (Abingdon: Routledge, 2003), p. 89.
18. Gillian Beer, *Open Fields: Science in Cultural Encounter* (Oxford: Oxford University Press, 1996, repr. 2006), p. 178.
19. David Hillman and Ulrika Maude, 'Introduction', *The Cambridge Companion to the Body in Literature*, eds. David Hillman and Ulrika Maude (Cambridge: Cambridge University Press, 2015), pp. 1–9 (p. 5).
20. John Berger, *Ways of Seeing* (London: Penguin, 1972, repr. 2008), p. 1.

21. Leo Braudy, *The Frenzy of Renown*, (1986) (New York: Vintage Books, 1997), p. 9; p. 12.
22. Stephanie M. Hilger, 'Introduction: Bridging the Divide Between Literature and Medicine', *New Directions in Literature and Medicine Studies*, ed. Stephanie M. Hilger (London: Palgrave Macmillan, 2017), pp. 1–12 (p. 2).
23. Marks, 'Reading the *DSM-5*', p. 165.
24. Jan G. van den Tweel and Clive R. Taylor, 'A brief history of pathology', *Virchows Archiv: European Journal of Pathology* 457 (2010), pp. 3–10 (p. 3).
25. 'Pathology, n.' *OED Online*, Oxford University Press, March 2023, www.oed.com/view/Entry/138805. [Accessed 16 March 2023].
26. Taylor, *Lords*, p. 7.
27. Matthew Schneider-Mayerson, 'Popular Fiction Studies: The Advantages of a New Field', *Studies in Popular Culture*, 33:1 (Fall 2010), pp. 21–35 (p. 22).
28. Victor E. Neuburg, *Popular Literature: A History and Guide from the Beginning of Printing to the Year 1897* (London: The Woburn Press, 1977), p. 12.
29. Emma Liggins and Daniel Duffy, 'Introduction', *Feminist Readings of Victorian Popular Texts: Divergent Femininities*, eds. Emma Liggins and Daniel Duffy (Aldershot: Ashgate, 2001), pp. xiii–xxiv (p. xiii).
30. Richard Faber, *Proper Stations: Class in Victorian Fiction* (London: Faber and Faber, 1971), pp. 12–13.
31. Len Platt, *Aristocracies of Fiction: The Idea of Aristocracy in Late-Nineteenth-Century and Early-Twentieth-Century Literary Culture* (Wesport, Connecticut: Greenwood Press, 2001), p. 26.
32. Ludwig Wittgenstein, *Philosophical Investigations*, trans. G.E.M. Anscombe (1953) 3rd ed. (1967) (Oxford: Basil Blackwell, repr. 1986).
33. Sianne Ngai, 'On Cruel Optimism', *Social Text Online* (15 Jan 2013): https://socialtextjournal.org/periscope_article/on-cruel-optimism/ [accessed 2 Apr 2023].
34. Maurizio Meloni, *Political Biology: Science and Social Values in Human Heredity from Eugenics to Epigenetics* (London: Palgrave Macmillan, 2016), pp. 37–38.

Works Cited

'aristocracy, n.', *OED Online* (Oxford University Press, June 2015) http://www.oed.com/view/Entry/10753?redirectedFrom=aristocracy#eid [accessed 7 July 2015].

Beer, Gillian, *Open Fields: Science in Cultural Encounter* (Oxford: Oxford University Press, 1996, repr. 2006).

Berger, John, *Ways of Seeing* (London: Penguin, 1972, repr. 2008).
Bourdieu, Pierre, *Distinction: A Social Critique of the Judgement of Taste* (1979), trans. by Richard Nice (1984) (London: Routledge, 1984, repr. 1999).
Braudy, Leo, *The Frenzy of Renown*, (1986) (New York: Vintage Books, 1997).
Bulwer-Lytton, Sir Edward, *England and the English*, 5th ed. (Paris: Baudry's European Library, 1836).
Cannadine, David, *Aspects of Aristocracy: Grandeur and Decline in Modern Britain* (London: Yale University Press, 1999).
Castronovo, David, *The English Gentleman: Images and Ideals in Literature and Society* (New York: Frederick Ungar Publishing Company, 1987).
Elias, Norbert, *The Civilizing Process* (1939), trans. by Edmund Jephcott (1978), eds. Eric Dunning, John Goudsblom and Stephen Mennell (Oxford: Blackwell Publishers, 1994, repr. 2000).
Elias, Norbert. *The Court Society* (1969), trans. Edmund Jephcott (Dublin: University College Dublin Press, 1983; repr. 2006).
Faber, Richard, *Proper Stations: Class in Victorian Fiction* (London: Faber and Faber, 1971).
Foster, R.F., 'Tory Democracy and Political Elitism' in *Parliament and Community*, ed. by Art Cosgrove and J.I. McGuire (Belfast: Appletree Press, 1983), pp. 147–175.
Foucault, Michel, *The Birth of the Clinic: An Archaeology of Medical Perception* (1963), trans. A. M. Sheridan (1973) (Abingdon: Routledge, 2003).
Frazer, James, *The Golden Bough*, 1922 edition (London: Penguin, 1996).
Gallenga, Antonia, 'The Aristocrat of Italy', *The Foreign Quarterly Review*, 28:56 (January 1842), 362–397.
Gilmour, Robin, *The Idea of the Gentleman in the Victorian Novel* (London: George Allen & Unwin, 1981).
Hilger, Stephanie M., 'Introduction: Bridging the Divide Between Literature and Medicine', *New Directions in Literature and Medicine Studies*, ed. Stephanie M. Hilger (London: Palgrave Macmillan, 2017), pp. 1–12.
Hillman, David, and Ulrika Maude, 'Introduction', *The Cambridge Companion to the Body in Literature*, eds. David Hillman and Ulrika Maude (Cambridge: Cambridge University Press, 2015), pp. 1–9.
Hyman, Gwen, *Making a Man: Gentlemanly Appetites in the Nineteenth-Century British Novel* (Athens, Ohio: Ohio University Press, 2009).
Kantorowicz, Ernst H., *The King's Two Bodies: A Study in medieval political Theology* (1957) (Princeton: Princeton University Press, 2016).
Lawrence James, *Aristocrats* (2009) (London: Abacus, 2010).
Lewis, Lady Theresa, *Dacre: A Novel*, ed. by The Countess of Morley, 3 vols (London: Longman, Rees, Orme, Brown, Green & Longman, 1834).
Lieven, Dominic, *The Aristocracy in Europe, 1815–1914* (Houndmills, Basingstoke: Macmillan, 1992).

Liggins, Emma, and Daniel Duffy, 'Introduction', *Feminist Readings of Victorian Popular Texts: Divergent Femininities*, eds. Emma Liggins and Daniel Duffy (Aldershot: Ashgate, 2001), pp. xiii–xxiv.

Marks, Christine, 'Reading the *DSM-5* Through Literature: The Value of Subjective Knowing', *New Directions in Literature and Medicine Studies*, ed. Stephanie M. Hilger (London: Palgrave Macmillan, 2017), pp. 165–79.

Meloni, Maurizio, *Political Biology: Science and Social Values in Human Heredity from Eugenics to Epigenetics* (London: Palgrave Macmillan, 2016).

Mulvey-Roberts, Marie, 'Introduction' in *Cheveley, or The Man of Honour* by Rosina Bulwer Lytton (1839), ed. by Marie Mulvey-Roberts, *Silver Fork Novels, 1826–1841*, 6 vols, series ed. by Harried Devine Jump (London: Pickering & Chatto, 2005), V, pp. ix–xxvii.

Neuburg, Victor E., *Popular Literature: A History and Guide from the Beginning of Printing to the Year 1897* (London: The Woburn Press, 1977).

Newby, Jennifer, *Women's Lives: Researching Women's Social History 1800–1939* (Barnsley: Pen & Sword, 2011).

Ngai, Sianne, 'On Cruel Optimism', *Social Text Online* (15 Jan 2013): https://socialtextjournal.org/periscope_article/on-cruel-optimism/ [accessed 2 Apr 2023].

'Pathology, n.' *OED Online*, Oxford University Press, March 2023, www.oed.com/view/Entry/138805 [Accessed 16 March 2023].

Platt, Len, *Aristocracies of Fiction: The Idea of Aristocracy in Late-Nineteenth-Century and Early-Twentieth-Century Literary Culture* (Wesport, Connecticut: Greenwood Press, 2001).

Schneider-Mayerson, Matthew, 'Popular Fiction Studies: The Advantages of a New Field', *Studies in Popular Culture*, 33:1 (Fall 2010), pp. 21–35.

Smith, D.A., 'Lewis, Lady (Maria) Theresa (1803–1865)', in *Oxford Dictionary of National Biography* (Oxford: Oxford University Press, 2004), Online edn., October 2006 http://www.oxforddnb.com/view/article/16595 [accessed 13 January 2015].

Taylor, Antony, *Lords of Misrule: Hostility to Aristocracy in Late Nineteenth- and Early Twentieth-Century Britain* (Houndmills, Basingstoke: Palgrave Macmillan, 2004).

Tillyard, Stella, *Aristocrats: Caroline, Emily, Louisa, and Sarah Lennox, 1740–1832* (London: Vintage, 1995).

van den Tweel, Jan G., and Clive R. Taylor, 'A brief history of pathology', *Virchows Archiv: European Journal of Pathology* 457 (2010), pp. 3–10.

Wittgenstein, Ludwig, *Philosophical Investigations*, trans. G.E.M. Anscombe (1953) 3rd ed. (1967) (Oxford: Basil Blackwell, repr. 1986).

CHAPTER 2

Fashionable Illness: Consumerism, Medicine, and Class in the Silver Fork Novels

> She went into Court, and testified publicly her faith in St. John Long: she swore by Doctor Buchan, she took quantities of Gambouge's Universal Medicine, and whole boxfuls of Parr's Life Pills. She has cured a multiplicity of headaches by Squinstone's Eyesnuff; she wore a picture of Hahnemann in her bracelet and a lock of Priessnitz's hair in a brooch.[1]

Introduction

In William Makepeace Thackeray's *Vanity Fair* (1848), Mrs Bute Crawley (one the novel's many socially aspirant characters) asserts, 'And as for *my* health, what matters it? I give it cheerfully, sir. I sacrifice it at the altar of my duty.'[2] The duty to which she refers is ostensibly the physical care of her husband's wealthy, ailing aunt. But within the parameters of Thackeray's satire, her 'duty' simultaneously broadens out beyond the family circle and shrinks to exclude it: Mrs Bute Crawley means to sacrifice her own health not out of any familial affection or obligation, but rather to forward her individual prosperity and to conform to the health tropes and medical markers of the upper classes.

Thackeray's novel—and its commentary on class, wealth, and health—operates in the confines of a nuanced and highly specific literary history. Though it is apparent through the narrator's commentary that *Vanity Fair*

A. Boucher, *Science, Medicine, and Aristocratic Lineage in Victorian Popular Fiction*, Palgrave Studies in Literature, Science and Medicine, https://doi.org/10.1007/978-3-031-41141-0_2

is a parody—even bordering on farce—it is likely unclear to most modern readers (and perhaps even to nineteenth-century readers only a couple of decades after the novel's publication) exactly *what* Thackeray is parodying. *Vanity Fair* was one of the very last, and certainly the best remembered, of the 'silver fork' novels; and Thackeray's scathing comedy, which was published as the genre's popularity waned, is often credited with killing off the genre entirely—a rather apt death for a genre which was obsessed with ideas of health, death, and imitative emulation.[3] Although the silver fork novels would have undoubtedly declined to the point of irrelevance without Thackeray's assistance, it is *Vanity Fair*'s protagonist, the coldly calculating and socially ambitious governess Becky Sharp, who hastened their already swift decline. Becky Sharp holds an uncomfortable mirror up to the silver fork's targeted readership, who often used the novels' tales of high life to attempt to facilitate their own social ascent.[4] And, like Thackeray's Mrs Bute Crawley, social ascent often went hand-in-hand with changing one's own medical health and physical identity. Working against Susan Sontag's assertion in her seminal treatise, illness—in the silver fork novels, at least—is nothing *but* metaphor.[5]

What follows is an examination of an early and nearly forgotten genre in nineteenth-century popular fiction, its circular conversation with medical rhetoric of the day (especially medical rhetoric rooted in classed issues), and three in-depth explorations of prominent silver fork novels which will serve as case studies for this chapter. I argue that despite—or perhaps because of—the perceived banality of the genre, silver fork novels were able to capture, reinforce, and problematise the complex portraiture of class as a medical issue in the first few decades of the nineteenth century in which illness (or at least the 'correct' illness) was considered a sign of good breeding. What's more, the modish nature of the novels, both in form and in content, help to trace the genre's highly integrated fluctuations of fashion, medicine, and identity politics during a time of considerable class reorientation and scientific advancement. As Sumiao Li argues about our understanding of fashion and class, 'often times, fashion in existing scholarship is simply understood as stylistic change in dressing and living, i.e. as part of consumerism and consumption. Frequently missing is the strong nineteenth-century sense of fashion as what rules Society [...] that special arena or system for norm-producing social activities.'[6] Despite their debatably deserved reputation for being shallow, the silver fork novels provide modern scholars with a vibrant window into an under-examined segment of early nineteenth-century life. Their discourse was so resonant that it shaped attitudes about class and medicine through the remainder of the nineteenth century, the clichés of which still exist in media today. The silver fork novels are a therefore a compelling place to begin an exploration of attitudes towards class and pathology in genre fiction of the long nineteenth century, the effects of which will reverberate over the course of the four subsequent chapters.

There are hundreds of silver fork novels from which to choose, and most of them contain at least some small element of engagement with the medical field. The novels I use as my case studies—Sir Edward Bulwer-Lytton's *Godolphin* (1833), Rosina Bulwer Lytton's *Cheveley, or a Man of Honour* (1839), and Catherine Gore's *Cecil* (1841)—were selected for three reasons. Firstly, these novels provide the reader with dynamic, intricate, and cogent readings of illness and class performativity. Secondly, these three novels also exemplify a surprising amount of range: although all written within a decade of each other, the novels were published at three distinct points in the genre's lifespan. They were also written by authors who had three different relationships to the upper classes: the first was born into it, the second ascended to it (through marriage to the first author), and the third was outside their ranks. Finally, these novels are more critically acclaimed, both at the time of their publication and today, than many in their cohort. While critical reception is not necessarily a reliable barometer for the significance of a novel's contribution to a field, nor its ability to capture the climate of the age, these texts possess a self-awareness and ebullience about their subject matter that makes for compelling close readings. Their popularity at the time also made these three texts epitomes of the genre, whose conventions were accepted and reproduced in other more formulaic (or outright facsimile) silver fork novels. These three case-study novels not only capture cultural paradigms surrounding class and illness, but helped to crystallise those paradigms through the novels' imitators.

Silver Fork Novels: An Overview of Genre

The silver fork novels, referred to sometimes as 'fashionable novels', enjoyed a robust popularity between the 1820s and mid-1840s before collapsing as a genre and fading almost entirely from the public consciousness. The appeal of the novels—or at least part of their appeal—is perhaps easier to ascertain and substantiate than other genre fiction before and since. Beyond purely the transmission of narrative, the silver fork novels also attempted to provide newly wealthy middle-class readers with an inside view to high society and to suggest to this target readership how they could use their disposable income to legitimise their new social positions.[7] Winifred Hughes argues that when the middle classes 'came to engage in the [...] process of self-definition, it was in terms of the established aristocratic standard that they were inevitably forced to measure their own system of values and code of conduct.'[8] The anxiety of class identity was potent enough for an anonymous contributor to *The New Monthly Magazine and Literary Journal*—a publication put out by silver fork publisher Henry Colburn, and edited by silver fork author Sir Edward Bulwer-Lytton—to comment in 1832: 'With us [English], there is a passion for inequality,—every one [*sic*] desires to be better than his fellows [yet the] creature who has been raised by his wealth, and his wealth alone, into a new sphere, feels uneasy and uncertain in his *position*.'[9] Political and economic hegemony was never a clear-cut issue in the silver fork novels. Even the increasingly threadbare and outdated aristocratic lifestyle—brought about by reduced financial circumstances in light of this class shift—carried with it its own romantic, stylish, and aspirational weight. Tamara S. Wagner writes in her 'Silver Fork Legacies: Sensationalizing Fashionable Fiction' that the trope of the 'shabby-genteel' seen in silver fork fiction and beyond 'was detailed with an ambiguous sympathy for its impoverished, disempowered representatives.'[10] Here we see the aristocracy—when not in their normal role as a glittering elite—placed in a dual position as continuing leaders of fashionable culture while also often idealised as something of an underdog. While the middle classes aspired towards aristocracy, the aristocracy could never (as Patrick Brantlinger argues) descend into the middle classes because 'the middle class as a whole is seen as progressively and upwardly mobile'.[11] The silver fork novels in many ways helped to bolster aristocratic authority through a new age by romanticising any lessening of authority within that age, and by introducing (or even falsifying) certain aspirational elements into the middle classes' growing class consciousness.

In particular, the genre resided at a production and marketing crossroads in terms of its readership's class practices. Or, rather, the genre savvily navigated, via its production and marketing, the class tensions of its readership's ambitions and socio-economic limitations. In her study of the history of Victorian literary businesses, Marrisa Joseph discusses how the combination of huge innovations in print technology and government involvement in education at the time of the silver fork novels' popularity 'consequently surged a demand for reading material'; the logical and, indeed, commonplace practice to meet this new demand was the serialisation of novels, as '[c]apital could be tied up in books for a long period of time, and profit could be realised more quickly within periodicals as they required "immediate and regular sales"'.[12] However, despite the greater flexibility and greater potential for quick profits, silver fork novels were typically *not* serialised. Rather, they were often published directly in two- or three-volume formats which were very popular within circulating libraries. The genre's often repetitive or reworked content (from many of the same authors) helped to supply regular material, even to the point of market oversaturation, for the increasingly literate masses. But publishers gave the (thematically appropriate) impression of readership gatekeeping by allowing readers only to access the novels in their final, expensive-to-purchase volume format, albeit under the far more affordable remit of a circulating library subscription. In doing so, silver fork publishers were able to create a genre that was seemingly exclusive and therefore could be framed as a seductive luxury, but in reality managed to reach a very wide middle-class readership.

The genre is notable for its focus on material culture, with each novel usually containing several lengthy passages listing fashionable items of dress, furniture, transportation, food, entertainment, and other markers of a socio-economic process that would come to be termed by Thorstein Veblen nearly a century later as 'conspicuous consumption'.[13] The tensions of material culture and class in these texts are noted by Clare Bainbridge in her introduction to Thomas Lister's 1826 novel *Granby*: 'silver fork novels already enjoy opposing the speed of the new macadamized roads and more efficient forms of stagecoach operation with the more leisurely methods of the past', and yet maintain their finger on the pulse of contemporary life by advocating for 'Stultz the tailor, Maradon the milliner, Gunter the *traiteur*-in-chief for aristocratic party-givers, Colinet as the provider of music for these same parties [and] Pasta and Catalani the singers and Keen the actor'.[14] Although the genre was not always known for its subtlety, as will be discussed below, its portrayals of material and medical culture evoke the byzantine, capricious, and even dissonant understanding needed to operate within the aristocratic apparatus. Portrayed as both atavistic and modern, both traditional and modish, both understated and sparkling, the genre's attitudes towards material culture illustrated the need for authors, characters, and aspirational middle-class readers alike to handle material goods with thought and deliberation.

More than material culture, these novels discuss the same need for prudence when it came to behaviours, rituals, tendencies, knowledge, and physiology—anticipating French theorist Pierre Bourdieu's concept of acquired cultural capital, *habitus*. Edward Copeland clarifies in his 2012 work on silver fork fiction that '"[f]ashion", the running trope of silver fork novels, is under review at all times, not as mindless "millinery", but as the drapery of power—in social occasions, political poses, dress, shopping, architecture, reading and manners'.[15] One could not simply buy one's way into the upper classes; one had to understand it, emulate it, and epitomise it flawlessly. And, as will be examined, aristocratic *habitus* in the silver fork novels was never so definitive, and yet so mercurial and fraught, as when these novels portray classed concepts of health and the body.

As many silver fork novelists were aristocrats themselves, or those who pretended a closer proximity to upper-class circles than they truly possessed, each aristocratically authored novel not only served as an entertainment and a self-help guide in one, but could also serve as a souvenir from an authentic celebrity life.[16] Some aristocrats found themselves impoverished and dependent upon the patronage of their middle-class readership; some, like Rosina Bulwer Lytton and her estranged husband Sir Edward Bulwer-Lytton (both silver fork novelists), used their public platform and the *roman à clef* nature of the books to defame or defend those in their social circle (including each other, during their bitter separation). The novels certainly tapped into the prurient interests of the public, but the allure of the *roman à clef* was more than a hunger for tawdry spectacle. Decoding the thinly veiled personalities seen in the novels—personalities that were sometimes used, satirised, and slandered by more than one author, or by a single author several times—was a part of the class education that the genre promised. The middle classes could more completely enter a world whose scandals were whispered behind closed doors, whose open secrets were only 'open' to the Upper Ten Thousand, and whose allegations appear without much context as blind items in *The Morning Post*. In short, silver fork novels would supposedly tell you what to buy, how to act, and would let you in on the joke.

The collective advice the genre furnishes is impressively comprehensive, if difficult to follow. Fashions changed so quickly that by the time a novel was published, its advice might no longer be *de rigueur*, assuming the advice was correct in the first place. Many middle-class authors who falsified their own backgrounds and aristocratic connections to publicise their books and boost sales did not always have accurate knowledge of the subjects upon which they wrote. Authentically aristocratic authors did not necessarily always paint the fullest picture of their class practices. Authors of both classes complicated these portrayals further through satire, sometimes very faintly, even during the early stages of the genre: before the literary conventions of the silver fork novels were themselves ripe for parody, authors occasionally ridiculed the aristocracy for their rigid rules and exclusivity, even as they stoked their readership's desire to understand those rules and be accepted into that exclusivity. Muddying matters further, novelists sometimes lambasted the class practices of the previous generation, making it more difficult to separate what was a paradigm of aristocratic existence from what was merely an outdated aristocratic practice. The silver fork novels, with their rapidly moving trends, interclass and intergenerational critiques, pseudo-biographical tendencies, and playful and sometimes reckless exposés, provide contemporary readers with remarkably rich portrait of popular culture in the first half of the nineteenth century. Despite the occasional and anomalous rhapsodies of critics (including one anonymous assertion in 1835 that the silver fork novels were to the nineteenth century what 'the biting sarcasm of Voltaire, and the burning eloquence of Rousseau' were to the eighteenth), it is only very recently that they've started to gain recognition as a subject worthy of academic study.[17]

Little-to-no critical attention was paid to silver fork novels as a genre in the nineteenth century, outside their individual and often caustic reviews. For most of the twentieth century, the only major work about fashionable novels was Matthew Whiting Rosa's 1936 *The Silver-Fork School: Novels of Fashion Preceding* Vanity Fair, in which Rosa speaks condescendingly of the genre, and uses silver fork novels only to contextualise *Vanity Fair*. Rosa's view of silver fork authors is likewise narrow, saying that of the dozens of authors who contributed to the genre, 'no more than eight merit attention—Theodore Hook, Thomas Henry Lister, [Robert] Plumer Ward, Lady Charlotte Bury, [Benjamin] Disraeli, [Sir Edward] Bulwer [-Lytton], Mrs. [Catherine] Gore, and Lady Blessington'.[18] Rosa's limitations on 'worthy' authors is apt, given the exclusionary and snobbish tone of much of the genre itself, but it does a disservice to our knowledge of the early nineteenth century to assume that less famous texts by less illustrious authors have nothing to contribute to the makeup and analysis of genre. Nor—more significantly—does Rosa's exclusive constraints allow that all of these texts taken *together* form a significant cultural artefact. I contend that all these texts form a map on which the transition of socio-economic power from the British aristocracy to the bourgeoisie is crystallised, along with the tensions, desires, and derisions felt and expressed by both classes who participated in its creation. Thankfully more recent critics have recognised the worth of such texts, the breadth and depth of which are invaluable to our modern understanding of fluctuations in class. Understandably, most current critical work largely focuses on the political, social, and economic and consumerist shifts represented most overtly in the genre.[19]

In part the novels' decline in popularity and correlating absence from scholastic consideration was the result of their decades-long, and near universal, bad reviews. And it is true that, more often than not, the novels were poorly written and extremely formulaic marriage plots or *buildungsromane* about a protagonist attempting to navigate the pitfalls of high society. Even the name 'silver fork' was conceived as an insult, originating in an 1827 article by William Hazlitt who, when reviewing the then-burgeoning genre, sarcastically noted that it did nothing more than teach people that the aristocracy ate their fish with two silver forks.[20] The novels were referenced negatively in *Jane Eyre*—they are read by Jane's shallow and unpleasant cousin Georgiana Reed—and provoked enough ire in philosopher Thomas Carlyle to merit an entire dismissive chapter, 'The Dandiacal Body', in his work *Sartor Resartus* (1836), in which he declared the genre unreadable.[21] Inexpensively produced and dedicated to transitory fashion, the novels—by design—aged very quickly and few novels merited more than one edition, with even the most popular books fading out of print after only a second or third edition.[22] A further consideration is that the novels' *roman à clef* nature and heavy intertextuality has place the genre under a cultural-literary bell jar. These novels exist about and within their own world, and can therefore seem alienating to readers for whom the references and in-jokes are obscured by both class boundaries and by historical distance.

The novels are in some ways a victim of their own success. By so masterfully encapsulating and epitomising the milieu of the early- to mid-nineteenth century, many of the texts quickly transitioned from ephemera to irrelevant. One of the few elements of these novels which escaped the genre's death, ironically, is their portrayal of aristocratic illness. The genre's collective interest in classed medicine not only resisted the effects of massive paradigm shifts but is still ubiquitous in twenty-first-century representations of aristocracy.

The Silver Fork Novels, Health, and Medicine

As previously discussed, recent scholarship—and, indeed, *most* scholarship—of the silver fork novels has largely focused on not only the socio-economic and political landscape depicted in the novels' pages, but also that which informed their target readership, authorship, and production. And it's true that the novels are probably most self-reflective and overt in their discussions of the marriage market, voting reform, and the shifts in class and wealth that have come so deeply to characterise our understanding of the first half of the nineteenth century. However, to focus only on these most conspicuous elements of the genre (especially when that genre provides commentary and guidance to its readers on hundreds of other subjects) would be a disservice to the genre's newly found critical interest. If the novels' overt purpose was to instruct readers in the ways they could transform themselves from middle to upper class, wealth was not the sole answer. In fact, many of the novels posit that wealth might even prove to be a hindrance to social climbers if they did not know how to harness it properly. Susan Sontag writes:

> With the new mobility (social and geographical) made possible in the eighteenth century, worth and station are not given; they must be asserted. They were asserted through new notions about clothes ('fashion') and new attitudes toward illness. Both clothes (the outer garment of the body) and illness (a kind of interior décor of the body) became tropes for new attitudes toward the self.[23]

Codings of the aristocracy go deeper than one's pocket and reside instead, as the novels often depict, in one's breeding. It was one's behaviour, both learned and intrinsic, both performed and physiognomic, that made one noble.

When discussing issues of class, a conversation about health is never far away. Throughout much of history, class was considered nearly as heritable as eye colour: a major consideration of class, especially in cultures where inheritance systems are rooted in primogeniture, is always going to be one of lineage, of blood, and of a 'correct' body born of a 'correct' body. Class implicitly brings up notions of heredity, medicine, and health, and these connections become especially potent in silver fork fiction: firstly, through the resonance of literature of sensibility; secondly, through the genre's use of medical issues as plot device; and finally through the cues that the silver fork publishing industry seemed to take from the late eighteenth- and early nineteenth-century British medical industry. Miriam Bailin writes in her text, *The Sickroom in Victorian Fiction: The Art of Being Ill*:

> Illness has always provided compelling images and sometimes plausible explanations for conditions of the spirit, the mind, the social body, and the body politic [...] illness can also register such desirable and diverse deviations as the Christian grace of affliction or the distinctiveness of a refined sensibility.[24]

Anticipating what we would now term a social constructionist stance on matters of the body, the silver fork novels weaponise the former naturalistic view of physical constitutions to maintain the exclusivity and unreachability of an elite group while at the same time fetishising physical constitutions as potential consumer goods or performative, appropriative class markers. What follows is a brief sketch of three socio-literary factors that most influenced the medicalised portrayal of aristocrats in this genre, culminating in close readings of the three case study texts.

The Cult of Sensibility

Although novels of sensibility, or 'sentimental novels', fell out of fashion roughly twenty-five years before the advent of silver fork fiction, their residual literary presence is keenly felt in the younger genre's articulation of the aristocracy as a pathologised institution and in its mockery of those attempting to 'perform' nobility. Sensibility, both as an emotional concept and as a literary genre, emerged as a reaction to the cool, self-possessed rationality of the Enlightenment in which the glorification of logic was supplanted by the glorification of emotion and sensation.[25] The delicacy of one's nerves and the refinement of one's sensibilities instantly triangulated the concept with greater issues of class and of medicalisation until the three issues became, for a time, inextricable. Michael Stolberg writes:

> The upper classes quickly adopted the new ideal as their own [...] the symptoms of a pathologically heightened nervous sensibility and irritability became a flexible means of self-fashioning, self-stylization, and self-dramatization. Those who put their vapors or nervous complaints on show in the public sphere, or indeed staged them with some degree of drama, could thereby give expression to their individuality, their moral qualities, and, at the same time, their distinguished social position.[26]

This is not to say such performances of upper-class ill health only developed in this period. Michel de Montaigne, writing in the sixteenth century, traces in his essay 'Not to Counterfeit Being Sick' a similar classed performance of delicacy as far back as the ancient Romans.[27] But the Cult of Sensibility's reach extended beyond performativity and into the realm of scientific and medical reality, with many celebrated physicians publishing their anti-Cartesian assertions that this sensitivity had a legitimate root in the bodies and minds of the upper classes. Biographer, diarist, and Scottish laird James Boswell justified his bad behaviour and its health consequences by asserting that his 'very miseries also marked [his] superiority', and even George III insisted that his madness was the result of her nervous sensibility.[28] The prominent eighteenth-century physician George Cheyne saw delicate health as a marker of both Englishness and luxury, saying in his *The English Malady* (1733):

> If *Nervous* Disorders are the Diseases of the Wealthy, the Voluptuous, and the Lazy [...] and are mostly produc'd, and always aggravated and increased, by *Luxury* and *Intemperance* [...] there needs no great Depth of Penetration to find out that *Temperance* and *Abstinence* is necessary towards their Cure.[29]

The later physician Thomas Beddoes, building in part from Cheyne's work, categorised the upper class through their delicacy and idleness; Beddoes in particular juxtaposed upper-class sensitivity to the different type of ill health seen in the working class brought on by their toil, and further juxtaposed it to what he considered to be the middle class's healthy athleticism.[30] Others viewed the conditions of health as not necessarily materially present in the body, but more the result of the habits, and even language, to which one was exposed: the eminent physician Sir William Osler, whose career spanned the last quarter of the nineteenth and first quarter of the twentieth century, suggested that

> to talk of diseases is not only ethically inappropriate, but potentially and dangerously transformative. In effect, to speak of illness is to replicate, linguistically, the process of its transmission from one subject to another. Once disease enters into language, its transmissive potential multiplies.[31]

Osler was no doubt at least partially referencing the effects of sensibility and the socio-medical zeitgeist of the late eighteenth century and its lingering effects throughout the century and its metaphorical—if not necessarily literal—transmission through ideology and cultural practice.

The ubiquity of these clichés in both medical literature and the arts coloured portrayals of the aristocracy well into the nineteenth century, and the trope of the medically delicate aristocrat pervades still today. Although sensibility as an artistic and emotional movement had faded by the advent of the silver fork genre, it gained new resonance in metaphorical and parodic realms around this time. Where once it had been fashionable for aristocrats to flaunt and perform illness and delicacy as a marker of their refinement, illness and delicacy now became for aristocrats in the early nineteenth century an inescapable and unwanted association. Historians of class have long noted the common perception of the slow death of the aristocracy over the nineteenth century, although David Cannadine complicates this in his 1999 *Aspects of Aristocracy* by stating that the British aristocracy presented itself as augustly stagnant 'while constantly evolving and developing'.[32] This portrait of august stagnation and even fashionable physical decline were perhaps slightly too effective; reified, complicated, and parodied in the silver fork novels, poor health has crystallised as a cliché ever since, despite the genre's frequent rebuttal of such modish deterioration.

The concept of 'fashionably' poor health was on the wane in the 1820s and 1830s: it was a passé reminder of the previous aristocratic generations' self-expression. Lord Byron, who features as a character in several silver fork novels (initially as an aspirational hero, but increasingly as a pastiche as the genre matured) was recorded wishing that he could die of consumption; when asked why by a friend, Byron responded: 'Because the ladies would all say, "Look at that poor Byron, how interesting he looks in dying."'[33] Even as early as 1810, when Byron stated his ill-advised wish, the idea of nobility-as-illness was breaking down: Byron didn't care if there was any actual connection between his fashionable allure and his illness, as long as the illness made him *seem* interesting. More crucially, due to the continuing rise of the middle classes, fashionably poor health signified the potential mortality of the upper class as a whole. British caricaturist Thomas Rowlandson's *The English Dance of Death* (1814–16) was a series of illustrated portraits in verse which satirised different social and personality types: in his chapter entitled 'The Genealogist', Rowlandson illustrates the socio-economic frailty of the aristocratic institution in more embodied terms:

> When debts unpaid assail'd his [an aristocrat's] gate,
> And his domains refus'd the weight
> Of mortgages, whose loud demands

Call'd for the sale of house and lands:
While the axe menac'd all the wood
Which round his noble mansion stood [...]
'Embarrass'd as I am,' said he,
'I'll not a noble beggar be,
'If means of honour can be found,
'To heal my fortune's wasting wound'. (118–19)[34]

The aristocracy was perceived as sick, and it was no longer just because they wished to be considered so. The Great Reform Act of 1832 began the slow process of voting reformation, taking the political hegemony away from the top 3 per cent of the wealthiest landowning males (usually peers) and putting it in the hands of the top 5 per cent; the number of voters would increase over the next seventy years, in fits and starts, to include all British men by 1900.[35] This same period, between the death of sensibility and the rise of the silver fork, saw a series of British royal succession crises; these crises began with the death of Princess Charlotte of Wales (only legitimate child of George IV) in 1817, continued with the death of George IV's next heir, his childless younger brother Prince Frederick in 1827, followed by George IV himself in 1830 without any legitimate issue, and finally the death of his brother and successor, William IV, who died in 1837 (also without any legitimate issue). The crises only abated in 1840, with the birth of Queen Victoria's first child, and came to a conclusive end over the next two decades as she safely gave birth to eight more children, all of whom survived until adulthood. One can also not underestimate the effect of the Corn Laws, enacted in 1815 and repealed in 1846, which illustrated the precariousness of the aristocracy's long-term financial solvency. The Corn Laws, which serve as neat bookends to the silver fork genre, were commented upon in the novels frequently.

Despite ill health's significantly more negative cultural connotation during the period of silver fork fiction, the genre illustrates the certain lingering glamour of sensibility; delicacy and illness are sometimes represented in the novels as aspirational goals for the middle classes. Most frequently ill health in these novels falls under the purview of *habitus*: it is the direct result of a fashionable and excessive lifestyle rather than an intrinsic component of those with upper-class lineage. In the Countess of Blessington's silver fork novel *The Confessions of an Elderly Gentleman* (1836), she writes, 'who, *without* taste, ever has the gout? and how few *with*, have ever escaped it!', while a character in Charlotte Trimmer

Moore's *Country Houses* (1832) refers to gout as 'a *gentleman's* disease'.[36] Gout, the result of excess uric acid in the bloodstream leading to a form of arthritis, was correctly understood centuries before the onset of silver fork fiction to be the result of diets rich in meat and alcohol; it was also common medical knowledge by this point that gout could be hereditary and specifically 'ran in good families'.[37] Whether it was an implicit health risk for those from 'good stock' or whether it was an avoidable 'disease of civilization, the bitter fruit of luxury', Roy Porter and George Sebastian Rousseau argue that gout was viewed by many as a 'superiority tax', the embodiment of a tension or an overlap between intrinsic or performative nature of being upper class.[38] For Blessington's speaker, who anticipates Bourdieu's *habitus*, taste and class are one—or at least they amount to the same physical markers.

Other novelists went even further, indicating the vagaries of fashionable trends and of class coding, especially as the genre continued into the 1840s. Some authors rejected sensibility outright, positing instead that aristocrats had, for years, spurned the notion as something so prolifically performed by the vulgar middle classes that it had lost all of its original aristocratic markers. The sign of an aristocrat was now one's robust good health, and the sign of wealthy *hoi polloi* was an ignorant adherence to an outdated form of class expression. For example, C.D. Burdett's 1828 silver fork novel, *At Home*, states, 'No person could enjoy more robust health than Lady Weldon did. But she had once dined where the fine lady of the mansion gloried in her *petite santé* [i.e. poor health].'[39] Tamara S. Wagner even traces in Catherine Gore fiction a mere nine years later Gore's characterisation of poor mental and physical health as a result of *middle-class* business and labour, instead of as a middle-class affectation, illustrating how truly out of date the concept had become.[40] That stylish ill health is still within recent memory for the now hale-and-hearty characters of the novel illustrates not only how performative these markers of class are, but also the collective cultural fugue necessary to perpetuate one's class status in the light of these changing markers.

The related notion of hypochondria was borne out of discourses of sensibility and class and, perhaps because of these changing mores on sensibility, hypochondria became a very bourgeois trait throughout the Romantic period. Michel Foucault famously argues in his *History of Sexuality* that whereas the aristocracy focused on notions of blood, bourgeois communities eventually focused on notions of health, writing, 'the aristocracy had also asserted the special character of its body, but this was in the form of *blood* [...]; the bourgeoisie on the contrary looked to [...] the health of its organism when it laid claim to a specific body.'[41] Both Roy Porter in his 1989 text *Health for Sale: Quackery in England 1660–1850* and George C. Grinnell argue that hypochondria eventually became a hall-mark of the middle classes and a growing 'symptom of social and political concerns such as matters of national self-identity, the economic comfort of the mercantile classes, and the unhealthiness of living in swelling cities.[42] Grinnell goes on to state that hypochondria

> primarily afflicted the bourgeoisie, so much so that it might not be possible to understand one without the other, and thus to speak of a nervous nation in the Romantic period is to acknowledge the dominance of the middle classes and their efforts to shape the nation in their own image as a collective body composed of relatively leisured citizens whose wealth made possible an age of medical consumerism.[43]

In this quotation one can see the interconnectivity of these issues: identity is shaped in the silver fork novels through wealth, through conspicuous consumption, through class markers, and through the body, blood and inheritance.

Disease and Death as Plot Devices

In addition to the remnants of sensibility and the intrinsically pathologised nature of class due to primogeniture and conceptions of breeding, the silver fork novels also engage with medicine and health in a much more formal way. The genre relies more heavily on dynamic plots than it does on rich character development or sophisticated tropes and these rote narratives were one of the biggest critiques of the genre. As such, disease and its cure are extremely common plot devices. Although hardly an unusual practice in genres that lean towards thrills and melodrama, illness and death are given an unconventional treatment in the silver fork novels in that they are often *positively* regarded by characters, authors, and readers alike, with medicine and cure sometimes serving as the antagonist in silver fork novels. The deaths of relatives, for example, are frequently sought by characters so that those characters may inherit rank and wealth. The illness or physical delicacy of a well-placed relative becomes an economic calculation in the marriage market. In Catherine Gore's *Cecil*, which will be explored in depth below, the reader is provided with a mock guide explaining how a young lady should go about making an advantageous marriage; the guide tells us that she should specifically target a 'peer, or a baronet, with a sufficient rent-roll [or] the eldest son of a peer or baronet, *whose father does not enjoy particularly good health*'.[44] An ill-timed recovery or cure may therefore serve as antagonistic forces that foil marriages, the repayment of debts, and social mobility, whereas illness or pathology is often a catalyst for progression. In Letitia Landon's 1831 silver fork novel *Romance and Reality* one character advises another that, 'All a young lady should pray for, is a severe lingering fit of illness, to impress upon her debating lover a just feminine valuation;—fevers and agues are the best stepping stones to the hymeneal altar', creating perhaps one of the stranger and yet more positive representations in the genre of physical deterioration as social progress.[45]

Given the cult of sensibility's continuing embedded nature within the aristocratic apparatus, it is unsurprising that pathology serves as an easy shorthand device to express character development—again with illness serving a generally beneficial role for the protagonist. A growing or shrinking susceptibility to certain illnesses, be they fevers or nerves or gout, can illustrate a character's ascent or descent on the social ladder. Edward Copeland argues, 'In silver fork novels, movement is what the reader watches for. Identity is defined from the body's space outwards.'[46] In the anonymously published *High Life* (1827), a character notes the inverse relationship between health and status: 'Georgiana certainly looks healthier than she did when she came from London [...] here she has the advantage of bathing, regular hours, and exercise.' This is a mixed blessing, as her time in the country and subsequent haleness signify a social regression.[47] Health here is not directly antagonistic, but it is a manifestation of other forces that are.

Charlotte Trimmer Moore takes a near-apocalyptic view of health and medical progress in her *Country Houses* (1832), writing:

> It has been asserted by Malthus [...] that if there were not contagious diseases to check population, the world must be over-run with inhabitants [...] but, since vaccination has been introduced, there would be no hope of human beings not at last devouring each other, if it was not for a disease sprung up in the higher classes of society, and from them, as a matter of course, likely to spread lower and lower. This extraordinary disease, unlike the plague, or small-pox, chiefly affected the *imagination* [...] In short, *to be bored* is so very dreadful, there is nothing one can encounter, *short of hanging*, that is not preferable.[48]

Here it is the particular brand of aristocratic ill health that acts as saviour of all mankind—paradoxically preserving life on a grander scale by destroying certain facets of life at an individual level. The *habitus* surrounding sensibility and its eventual metaphorical infection of the larger, lower classes is seen as offsetting the progress made over the increasing robustness of the lower orders, whose health inevitably brings death.

The silver fork novels have an occasional tendency to portray health as a zero-sum-game, in which one character's health often comes at the expense of another's. Matching the rote nature of the novels (as well as reproducing the cycles of high society with its perceived quick turnover of members for inheritance and the marriage market), silver fork fiction espouses a certain antipathy towards medicine and good health as a road block to narrative progression and the fulfilment of its characters. This is both unusual and formulaic: the genre's attitude towards health-as-antagonist is not at simplistic as it first appears. Health, in many cases, needs to be vilified to create mid-novel tension for a more satisfactory resolution, but also to create self-reflective commentary. The novels comment on the ineffectiveness of the admittedly alluring primogeniture system that rests upon the restrictive and privileged factor of inheritance. More complexly, however, these novels and their treatment of health as villain create a commentary on the consumption of both aristocrats as celebrities and luxury goods, and the consumption of silver fork novels themselves. For a genre and culture that glorifies youth, beauty, wealth, and the tensions of the marriage plot, it is therefore desirable that aristocratic lives (both fictional and real) have a relatively rapid turnover. The quicker and more dynamic the life, the sooner an audience will again get to partake in the entertainment provided by another similar plot, with new space made for the children of the previous protagonists or other newly wealthy, young, and compelling participants. Readers are led by silver fork authors to cheer the on the deaths of characters who stand obstruct the financial and romantic paths of the protagonists and then, conversely, to champion the long life and happiness of those same characters who now stand in the way of a younger crop of aristocrats about to embark on the same plight. Seen in this light, the fraught and multifaceted portrayals of health transform slowly from 'villain' to a gentler concept of 'conflict'.

Business Models

The silver fork genre's connection to medical culture and rhetoric goes beyond the content and style of the literature. The business model of silver fork publishing seemed in many ways to be based on, or take its cues from, late eighteenth- and early nineteenth-century healthcare business strategies—especially, Marrisa Joseph notes, as the literary industry professionalised and standardised over the course of the nineteenth century.[49] More than mere professional standardisation bringing the literary industry out of its 'quasiprofessional occupational group' status and further into the realm of the 'traditional professions' (the Church, the army, medicine, and law), the increased professionalism of the nineteenth-century publishing industry was, rather aptly, 'an attempt to act as a barrier to entry for newcomers wishing to enter "the sacred company"', with their own exclusive clubs doubling as both social and professional networks.[50] While I don't argue any direct causation between various nineteenth-century healthcare fields and silver fork publishing, there was certainly a correlation or overlap of some practices which are worth noting—most particularly both industries' conceptual tensions between professionalism and elitism, their social and occupational remit, and their role as both facilitators *and* embodiments of classed celebrity.

The silver fork genre operated in a period of mass consumerism and, as Cheryl A. Wilson says, 'the novels primarily educated readers into becoming middle-class consumers.'[51] One especial area of consumerism during this period—just as today—was healthcare, including amongst other things the marketing and sale of faddish medicines and devices, services from high-profile physicians, and holidays to stylish health resorts and spas. Roy Porter writes that '[c]ommercial medicine cashed in on the prestige symbols of the times, aiming to make medicine smart, intriguing, and even fun'.[52] He continues:

> In this age when people were said to be 'sick by way of amusement' and fashion supposedly dictated people's choice of illnesses, doctors, and medicines, it is not surprising that sophisticated medical service sector emerged, equipped with manufacturing depots, warehouses, nation-wide wholesale and retail marketing facilities, and mass advertising to care for, or cash in on these need.[53]

The rise of commodity culture in this period, triangulated with socio-economic shifts and the effects of sensibility, led to a breeding ground in which classed medicine and fetishised pathology were prevalent to the point of inescapability. Both industries, healthcare and silver fork publishing, encouraged certain demographics through often manipulative means to purchase new and ultimately incomplete and unfulfilling goods and services—incomplete and unfulfilling, since it is unlikely anyone could purchase *enough* good health or social advancement to sate individual ambitions, nor would such states last without continuing effort and investment.

The circumstances of the medical marketplace around the turn of the nineteenth century led to the rise of quackery. Satirised by Rowlandson, we can see the porousness between class boundaries and between truth and fiction is only due to the authority granted by fashion:

> —Do, for a curious Moment pop
> Your head into that splendid shop,
> Where gilt vases, in a row,
> Form such a gay and motley show;
> With labell'd bottles to explain
> The compounds which they ne'er contain:
> Where Juleps, Anodynes and Pills
> Are seen prepar'd to cure all ills
> That do infect the human frame,
> Whate'er their nature or their name.[54]

Quackery was a classed issue in itself, fuelled on by aristocratic patrons and by medical mal-practitioners who became aristocrats themselves through knighthoods, like the sham oculist William Read who was knighted for his royal (dis)service to Queen Anne.[55] Genice Ngg traces the popular connection between quacks and aristocrats (at least in literature), going as far back as the early seventeenth century, in works like Ben Jonson's *Volpone* (1607), the Earl of Rochester's 'Alexander Bendo's Bill' (1676), and Aphra Behn's *The Second Part of the Rover* (1681), although these connections would only grow through the eighteenth century.[56] Even Charles Dickens used aristocratic quacks as part of his historical set dressing in his 1859 novel, *A Tale of Two Cities*, showcasing doctors of the 1780s 'who made great fortunes out of dainty remedies for imaginary disorders that never existed [in] their courtly patients'.[57] Miriam Bailin writes that 'the nineteenth-century upper and middle classes were treated at home, sometimes, though not always, by a doctor whose professional qualifications resides as much in his relations to the family, standing in the community, and personal charm as in his technical skill and knowledge'.[58]

Medical practitioners—both legitimate and fraudulent—were aware of social distinction and its correlating rewards available to those who managed to permeate class boundaries.[59] And, as many doctors of the period came from middle-class families (joining the medical field in a professional capacity was an unlikely scenario for upper-class men), being knighted served as yet another version of the improbable middle-class aspirations fostered by the silver fork novels. What's more, medical practitioners who *did* become the purveyors of elite services not only often became members of the elite themselves (despite the labour they provided to their now supposed equals), but also became barometers of status themselves. In an ouroboros of namechecking and conspicuous consumption, medical practitioners gained their elite status through services provided to aristocratic patients, who themselves gained further status through purchasing the services of those elite medical practitioners.

To enter into this echo chamber of fame and status, physicians and the creators of patented medicines often falsely advertised their connection to aristocratic circles to give an initial weight and authority to their products:

> Empirics selected their brand-names to chime with the fashionable, elitist, progressive aspirations of Enlightenment high society, evoking the reputations of top scientists, the cosmopolitanism of exotic wisdom, and philanthropic associations of ecumenical religion [...] Above all, quack medicines were commonly named (legitimately or not) after famous orthodox physicians, such as [Richard] Mead, [Sir Hans] Sloane, or [John] Fothergill.[60]

Henry Colburn, easily the most prolific of all silver fork publishers, also used this technique, trumpeting, exaggerating, or falsifying his connections to aristocratic authors in order to promote the sale of books.[61] And, indeed, many of Colburn's authors had genuine connections to the aristocracy, including Lady Caroline Lamb, Robert Plumer Ward, Lady Morgan, Thomas Henry Lister, Lady Charlotte Campbell Bury, George Payne Rainsford James, Lady Theresa Lewis, Sir Edward Bulwer-Lytton, Rosina, Lady Lytton, Lord Normanby, and Arabella Jane Sullivan.

Others—like 'Lady Humdrum' (Mrs Alexander Blair), Letitia Elizabeth Landon, Theodore Hook, Harriet Pigott, and Benjamin Disraeli (who, at least before his political career took off in 1837, had few grand connections and erroneously put about that his father was descended from notable Continental families)—had their connections either fully invented or heavily embroidered. In her introduction to Letitia Landon's silver fork novel, *Romance and Reality* (1831), Cynthia Lawford writes:

> By merely agreeing to write a fashionable novel for [Henry] Colburn and [his sometimes-publishing partner Richard] Bentley, Landon was forced to assume a false position. She had to present herself as more *au fait* on fashionable matters than she really was. Readers of silver fork fiction expected such expertise of their novelists, and reviewers might have declared her an imposter if she failed to convince, as they had done other novelists.[62]

Colburn's advertising strategies regarding the authenticity of his authors and readers was so successful that the 'going opinion of literary historians until only recently has been that readers of Colburn's novels of fashionable life were comfortably well-off members of society—exclusives themselves'.[63] The perceived limited access to the novels and restricted *entrée* to the world of *ton* was, as Copeland pointed out, as fictional and yet as omnipresent as the medicalisation of aristocrats in popular culture itself.

Whether he received inspiration from the medical community or elsewhere, Colburn's particular type of branding was a notorious and frequently criticised practice duplicated in medical circles and, as will be discussed, lambasted in the silver fork novels themselves. Thomas Percival in 1803 heartily denounced quackish practices in his *Medical Ethics*, saying that they were

> disgraceful to the profession, injurious to health, and often destructive even of life' and that the dispensation of 'secret [a] *nostrum* [...] is inconsistent with the beneficence and professional liberality. And if mystery alone give it [a medicine] value and importance, such craft implies either disgraceful ignorance, or fraudulent avarice.[64]

The magnitude of quackery led to an early nineteenth-century desire for greater regulation, regulation that was spurred on and organised by *The Lancet*, whose editor at the time was the radical pro-Charist reformer and surgeon Thomas Wakley. The journal featured, for decades, regular columns and opinion pieces railing against quackery in all its forms. In its first year, the journal ran a column entitled 'Compositions of Quack Medicines' which listed the ingredients of unreliable medications. One supposed remedy for deafness was revealed by this column to be little more than almond oil infused with garlic; another, this time for 'hooping [*sic*] cough', was mostly clove-scented olive oil.[65] This desire for reform prompted, over the course of the century, the Apothecaries' Act (1815), the Medical Registration Act (1858), the Pharmacy Act of 1868. It also led to the founding of several societies to help in 'identifying and combating quackery [such as] the British Medical Association (1832), Pharmaceutical Society of Great Britain (later Royal Pharmaceutical Society) (1841), and British Dental Association (188)'.[66]

Quackery's relationship to fashion, celebrity culture, and, above all, the aristocracy led rather organically to its inclusion as a topic of ridicule in silver fork novels. Despite the novels' promotion of illness via the cult of sensibility, the faddish trends and products seen it the genre's pages were treated with the same half-sincerity and half-satire as other aristocratic practices depicted in the novels. Charlotte Trimmer Moore's 1832 silver fork novel *Country Houses* navigates this line of satire well, with one of the aristocratic characters advising:

> I am one of those unfortunate persons [...] who are dyspeptic [...] indeed, I take charcoal before I dare touch any thing [*sic*] [...] it is the most fashionable medicine now, *pure carbon*, and [...] I believe it has saved the life of my nephew, at Malta: they give it with great success in the fever prevalent there, and at Paris they take it for every disease under heaven. The fine ladies take it constantly, and carry it in their reticules in the form of *bon bons*.[67]

The medical advice depicted here is given out with rather dangerous certainty, by someone who is clearly a non-expert—the aristocratic character speaking—based only on hearsay evidence and focused more on the product's status as an accessory rather than as a legitimate medical product. Whether intentionally ironic or not, Trimmer Moore's criticism of quackery could easily apply to her own fashionable product for sale, especially given that Trimer Moore has no known connection to high society. The advice she espouses in her own text, ostensibly for the betterment of the life of its consumers, is based on the same hearsay evidence or plausible-sounding invented logic as the medical malpractitioners caricatured in her narrative.

Silver Fork Novels: Three Case Studies

In the below section, I examine three silver fork novels in depth as case studies of the pathologised nature of the genre: Edward Bulwer-Lytton's *Godolphin* (1833), Rosina Bulwer Lytton's *Cheveley* (1839), and Catherine Gore's *Cecil* (1841). In my analysis of these three silver fork novels, any commentary on the veracity of a character's claims to illness will be limited and will be restricted to the texts' overt depiction of a particular medical issue as an authentic marker of class, or as a conscious exhibition or mimicry of aristocratic *habitus*. Even where a character's medical complaint is overtly stated to be performative—as in the case in Lady Charlotte Campbell Bury's character Lady Ellersby in *The Exclusives* (1830), who states that 'One don't look well when one faints—that is to say, *really* faints [...] it is surely best to avoid doing so'—I will make no claims about the pathology of such a performance.[68] To bring up Munchausen's Syndrome, psychosomatic symptoms, or hypochondria would be—with the exception of the latter—anachronistic and would miss the point of the texts' importance to class-based medical and scientific discourse. Hypochondria as we know it today (by its new diagnostic terms 'somatic symptom disorder' or 'illness anxiety disorder') is a mental disorder that produces '[p]ersistently high level[s] of anxiety about health' in sufferers who believe they detect in themselves the symptoms of (usually) serious illnesses, or have 'disproportionate and persistent thoughts about the seriousness' of the symptoms they do possess.[69] The understanding of hypochondria as a somatic psychological disorder had certainly solidified by the time of silver fork novels, and is even referenced overtly a few of them, including Edward Bulwer-Lytton's *Godolphin*, explored below.[70] However, this chapter is less concerned with the 'realness' versus 'faking', or the mental and psychological origins of an illness and its physical symptoms. Elaine Scarry argues in her work on pain and the body, that to 'have pain is to have *certainty*; to hear about pain is to have *doubt*'; elements of doubt or performance are often so central to the silver fork genre's conceptualisation of illness that they go without saying.[71] The authenticity of an illness is not as significant as the 'authentic' class behaviour in response to that illness. Rather, what is germane to is chapter is the assimilation of the general culture of illness, medical issues, and cures into fashionable lifestyles, both as parts of a luxury market and as objects of aspiration or ridicule.

Godolphin

Sir Edward Bulwer-Lytton's 1833 novel *Godolphin* is perhaps one of the better-remembered silver fork novels due in part to its status as a serious class commentary, as well as for its bizarre inclusion of a supernatural segment in the middle of the novel. Its adherence to silver fork structures and tropes at the very height of the genre's popularity seems incidental on the part of Bulwer-Lytton; his discussions of man's celestial and spiritual nature include enough material on aristocratic practices to produce, perhaps accidentally, an exemplar of silver fork fiction. Despite the oddness of Bulwer-Lytton's more philosophical and occult sections, they do serve to embed his text into the generic conventions of silver fork fiction by contemplating certain physiological or medical issues, including the fashionable concepts of magnetism and mesmerism. Harriet Devine Jump writes in her introduction to *Godolphin*:

> [I]t was Bulwer's conviction that it was the duty of the artist 'to suggest the sublime element of human existence by disclosing the ideal or transcendental within the real or concrete' [...] Among other early reading, he certainly knew *Anthropo-metamorphosis: Man-transformed; or, The Artificial Changeling*.[72]

Anthropo-metamorphosis (1650), written by physician John Bulwer (Bulwer-Lytton's distant relative), was a very early work of anthropology and body theory which focused, much as Bulwer-Lytton's much later novel, on the cultural significance of modification, enhancements, and deformations of the body. *Anthropo-metamorphosis*'s frequent readings as a veiled piece of political theory reinforce Bulwer-Lytton's own later literary intermingling of politics and the body through the lens of class-based medical practice and identity.

Godolphin was not Bulwer-Lytton's first silver fork novel, nor even his most influential. Bulwer-Lytton began his silver fork career five years earlier with *Pelham; or, the Adventures of a Gentleman* (1828) which solidified the genre by building upon the conventions set forth by Theodore Hook's *Sayings and Doings* (1824–28), Robert Plumer Ward's *Tremaine, or the man of refinement* (1825), Benjamin Disraeli's *Vivian Grey* (1826), and T.H. Lister's *Granby* (1826). A prolific novelist, Bulwer-Lytton wrote three other novels in the intervening years which have been classed as silver fork novels only to varying degrees by critics (and even Bulwer-Lytton himself, who insisted that the first of these was a metaphysical work): *The Disowned* (1829), *Devereaux* (1829), and *Paul Clifford* (1830).[73] It was Bulwer-Lytton himself, in his introduction to *Godolphin*, who stated that '*Pelham and Godolphin* are the only ones which take their absolute groundwork in what is called "The Fashionable World"'.[74]

Despite Bulwer-Lytton's pairing of the two novels, and despite the formulae of the genre, *Pelham* and *Godolphin* are extremely divergent in style and outlook. *Pelham*, which greatly increased Bulwer-Lytton's stature as a novelist, gently ribbed aristocratic conventions through the figure of the high-society dandy. *Godolphin*, which alternates between far greater idealistic and cynical extremes, is a doomed romance about the infiltration and assimilation of lower-class protagonists into the aristocracy; *Godolphin* also makes considerably more use of medical discourse in relation to class. The drastic shift in tone by the same author, writing within the same genre, with only a five-year gap between the two texts, could be in part due to the crystallisation of silver fork tropes and reader expectations at this period, or potentially due to Bulwer-Lytton's changing personal experience with class boundaries: in 1827, the year before he published *Pelham*, he married Rosina Doyle Wheeler, an Irish beauty from a landowning family who was nevertheless beneath Bulwer-Lytton's station. Their marriage was an unhappy one and they informally separated six years later, the year he published *Godolphin*. The beautiful but cold and misanthropic heroine of *Godolphin* bears a resemblance to Bulwer-Lytton's estranged wife and the eventual marriage between the novel's hero and heroine is similarly rocky (albeit nowhere near as sensational as the marriage between the two authors).

Godolphin tells the story of the romance between the two non-aristocratic leads, Percy Godolphin and the headstrong and capable Constance Vernon. Constance initially rejects Percy in order to fulfil her father's deathbed wish that she marry into the aristocracy to exact revenge specifically upon her father's political rivals and to break down the aristocratic institution from within. She marries Lord Erpingham and becomes a leader of high society with ease, shaping the fashions of the day and greatly influencing the outcome of Parliamentary votes. Despite her secret resistance to her aristocratic circle, Constance becomes enmeshed in its practices and tropes, and her health correspondingly weakens. Percy nurses his romantic rejection on the Continent, where he becomes the student of an astronomer, philosopher, and soothsayer named Volktman. Eventually Constance and Percy are reunited in marriage and are able to briefly enjoy their philosophical, financial, social, and political successes in London before Percy tragically dies, in accordance with one of Volktman's prophesies.

Godolphin opens, appropriately enough, on Constance's father's deathbed scene, catalysing the heroine to launch herself into the upper echelons of high society. This correlation of death with new beginnings, lively action, and class mobility is something that dogs many characters in silver fork fiction, but negatively so in *Godolphin*: if death prompts aristocratic action, aristocratic action only leads to further death. Constance's father, a middle-class political genius used and discarded by his former aristocratic friends, traces the roots of his ill health back to the aristocracy:

> I remember the day when I was ill after one of their rude debauches. Ill!—a sick headache—a fit of spleen—a spoiled lapdog's illness! Well: they wanted me that night to support one of their paltry measures—their Parliamentary measures; and I had a Prince feeling my pulse, and a Duke mixing my draught, and a dozen Earls sending their doctors to me. I was of use to them then.[75]

It is not merely that the rude debauches lead to Mr Vernon's ill health, both temporarily and long term, but that the active participation of high ranking members of society in medical culture leads him to further imprudent behaviour within that culture and, indeed, even a glamorisation of the abject and unpleasant. However, without the same social privileges and protections enjoyed by those of rank, Mr Vernon's ailments are ultimately not seen in the same light, nor treated with the same leniency that borders on approval for those in the *bon ton*:

> With my health my genius deserted me; I was no longer useful to my party; and when I was on a sick bed—you remember it, Constance—the bailiffs came and tore me away for a paltry debt—the value of one of those suppers the Prince used to beg me to give him.[76]

Further, the middle classes are so far removed from the decadent aristocratic practices that lead to illness and death that middle-class bodies do not even seem to be of the same species as aristocratic bodies—or at least their bodies operate under different natural laws and within different physical systems. Where little or nothing is said about the intrinsic health status of aristocrats and while their medical status is solely tied bad behaviour, middle-class bodies are represented as having intrinsic value and perhaps a greater class identity through good health. Mr Vernon's health becomes corrupted only when he blurs class lines; his health and class identity are spoiled irrevocably, and in equal measure. It is only after years of suffering in poverty and rejecting the interference of medical professionals—'No, child, no; I loathe, I detest the thought of [medical] help [...] Thank God you are my only nurse and attendant'—that Mr Vernon's physical form regains some of its original class markers, although it cannot overthrow the fatal effects of aristocratic contamination.[77] Bulwer-Lytton writes of Mr Vernon's wasted countenance:

> But what a countenance it still was! The broad, pale, lofty brow; the fine, straight, Grecian nose; the short, curved lip; the full, dimpled chin; the stamp of genius in every line and lineament;—these still defied disease, or rather borrowed from its very ghastliness a more impressive majesty.[78]

Here Bulwer-Lytton projects a dialogue, if not outright battle, between the two classes: Mr Vernon's value and identity is clearly marked not only through his beauty, but through his talent. It is his genius that is, or once was, on equal footing with the grandeur of an aristocratic title, and is also a testament to his class: he ascends or descends on personal merit and the strength of his labour, instead of the stagnation and security of those born into the elite. That his features defy 'or rather borrow from' aristocratic bad health serves as a microcosm of the novel as a whole: the middle classes rally against and reject the aristocracy, but the aristocracy is ultimately too powerful and engrained. Mr Vernon's physical form seems to misappropriate aristocratic ill health to better reify his own middle-class 'stamp of genius' before ultimately being killed by that ill health; so, too, does Constance infiltrate the aristocracy and manipulate elements of that class to work against itself, before her life is finally ruined. In both situations, the aristocracy and its tropes may smart briefly, but they ultimately carry on untouched.

Mr Vernon's physiognomic class struggle, in some ways, infects his daughter as well. The then thirteen-year-old Constance internalises both her father's successes and his failures within the class system, along with their corresponding medical crises. His alternating adherence to two different class groups embeds itself in his daughter's physical form: 'opposite the dying man now stood a girl who might have seen her thirteenth year. But her features—of an exceeding, and what may be termed, a regal beauty—were as fully developed as those of one who had told twice her years.'[79] Constance's beauty and physical advancement over her peers mirrors the beauty and middle-class genius seen in Mr Vernon's features. As with Mr Vernon's features, it is tainted by the threat of aristocratic physical decay: her beauty is qualified as 'regal', which then shades her development 'as those of one who had told twice her years', aging ahead of her time for better or worse. Constance's possession of a physically mature body is, in this moment, portrayed as beneficial to the matrimonial chances of one so young. The elite nature of this body, coupled with the rather startling assertion that a thirteen-year old child could appear to be twenty-six—not just a few years older, but nearly a decade and a half—indicates an unhealthy or unhelpful physical element coupled with aristocracy and rank. Constance's physical maturity places a significant burden on her unequal intellectual and emotional maturity, and this burden serves as the antagonistic force of the novel. Further, the taint of aristocracy within her highly advanced physical development leads the reader to wonder if Constance has fully lost these years of her life, and if her body is nearing death at twice the rate of her middle-class peers.

Mr Vernon, who revels in 'the deadly and dooming weight of my dying curse', does not fully realise that he is dooming his daughter's life as well.[80] Although Constance survives the whole narrative, she is surrounded by illness, death, and decay from the moment of his imploration:

> Swear that you will not marry a poor and powerless man, who can minister not to the ends of that solemn retribution I invoke! Swear that you will seek to marry from amongst the great; not through love, not through ambition, but through hate, and for revenge! You will seek to rise that you may humble those who have betrayed me! In the social walks of life you will delight to gall their vanities; in state-intrigues, you will embrace every measure that can bring them to their eternal downfall. For this great end you will pursue all means.[81]

By asking Constance firstly to follow his path and embed herself in the aristocracy, and then to mimic the aristocracy in its purest form by becoming an agent of death and destruction, Mr Vernon condemns his daughter and gives power to the very pathologised class group he is intent on eradicating. Constance, who marries Lord Erpingham a few years later in accordance with her father's wishes, undergoes a correlating physical deterioration as her social supremacy solidifies: 'When people have really nothing to do, they generally fall ill upon it; and at length, the rich colour grew faint upon Lady Erpingham's cheek; her form wasted; the physicians hinted a consumption, and recommended a warner clime.'[82] Constance, although considering herself as merely performing aristocracy at this point, is not here performing illness but is rather a legitimate victim of it. She has gone from being a middle-class woman masquerading as the member of another class to being a genuine member; her illness is based not on intrinsic class physiology but on behaviour and its effects. Despite her plans and savvy political machinations, Bulwer-Lytton emphasises the decadence of her life by saying she has '*really* nothing to do', and that that the pointlessness of and lack of meritocracy in primogeniture systems is the source of contagion, illness, and decrepitude.

Bulwer-Lytton's representation of aristocratic illness in this novel revolves around the breakdown of sensibility: in *Godolphin*, illness is undeniably fashionable, but only because it has come to be associated with the aristocracy through their class *practices*, and not through an intrinsic or in-born delicacy or susceptibility. The more Constance Vernon rallies against the institution of which she is now a part, the more she is celebrated and feted for her elusive and exclusive persona. In conjunction with her accidental epitomisation of the class she despises, her physical form begins to react in equally contradictory ways and as she wastes away physically, she becomes reified socially.

Constance's status in the aristocracy not only destroys her own health, but is also a source of contagion and death to nearly every character deeply involved in her life, but especially to her first husband (Lord Erpingham, who dies partway through the narrative) and her second husband—the novel's protagonist, Percy Godolphin. Her first husband dies as a direct consequence of her own misdiagnosed illness; taking her to fashionable resorts in Italy, he

> diverted the tedium of a foreign clime with a gentle ride. He met with a fall, and was brought home speechless. The loss of speech was not of great importance to his acquaintance; but he died that night, and the loss of his life was![83]

Although his death not the direct result of an illness, per se, it *is* in line with the narrator's earlier assertion about the dangers of having 'nothing to do'. Lord Erpingham's death stems twofold from illness-related privilege—his wife's boredom leads to her illness, which leads to his boredom and resulting death. The chipper, cavalier tone the narrator takes towards Lord Erpingham's death underscores how deeply integrated death, illness, and change are within the novels and within aristocratic society. For all the weight placed on lineage, primogeniture, marriage, and the reification of existence through sentimental suffering, life itself is portrayed as remarkably unimportant, especially if it occurs within fashionable parameters.

Shortly after Constance's widowhood, she and Percy Godolphin meet again in Italy and rekindle their early romance. Percy, also romantically involved with a peasant girl during his more wholesome, spiritual adventures on the Continent, is torn apart by guilt and a return to the aristocratic lifestyle he shrugged off early in the narrative:

> 'Ill!—ha, ha! I was never better; but I have just returned from a long journey: I have not touched food nor felt sleep for three days and nights. I!—ha, ha! no, I'm not ill;' and, with an eye bright with gathering delirium, Godolphin [...] stood arrested for one moment, and [...] burst suddenly into tears. That paroxysm saved his life; for days afterwards he was insensible.[84]

Here sensibility plays a role again, working as a safety valve to Godolphin's intraclass struggles; pained by the woman (and the class) to whom he should swear his allegiance, his physical form seems to make the choice for him and declare its implicit aristocratic sensibility through a sentimental 'paroxysm'. Bulwer-Lytton even goes so far as to give Percy a full class-based physical reset through his following days of insensibility, after which he aligns himself fully with Constance—ultimately to his death in the same manner as her first husband. The repetition of a fall from a horse is used by Bulwer-Lytton to show the cyclical nature of aristocratic illnesses and death, both rooted in behaviour. For example, both Percy and Constance, under their own remit, go to Italy to escape these aristocratic cycles and influences (he to escape his disappointment in the marriage market, she to escape the boredom-inducing sickness of British high society) only to get pulled back in to those same structures and patterns. It is lost on both of them that they choose a fashionable foreign locale, which reproduces their own world in an exotic microcosm.

These repetitions reinforce Bulwer-Lytton's stance on the far-reaching medical effects of aristocracy and its propensity for its own destruction. Several tertiary characters die after an upper-class indoctrination to gambling: Percy himself is declared early in the novel to 'look feverish [...] I don't wonder at it, you lost infernally [gambling] last night', while an acquaintance named Johnstone commits suicide after a bad night at cards, while yet another friend 'died after a hearty game of whist. He had just time to cry "four by honours," when death *trumped* him'.[85] As with the death of Lord Erpingham, the tone is always jovial; the behaviour and its consequences both expected by and trivial to all involved, except to the dead themselves. Constance's misdiagnosis in the middle of the novel—a supposed consumption which leads her to Italy—is a further pattern in which aristocrats become trapped. Bulwer-Lytton repeatedly classifies physicians as 'quacks and pretenders that have wronged all sciences', yet to whom aristocratic characters cannot help but return.[86] As these physicians are part of the system whose very identity is tied up in the concept of ill health, it is in their interest twofold *not* to cure it: firstly to ensure that their noble clients do not lose all traces of their class status through too robust health, and secondly because their economic success depends upon repeat custom. Bulwer-Lytton writes, that 'positive knowledge of pathology is what no physician [...] really possesses. No man cures us—the highest art is not to kill!'[87] In this way, physicians in this novel walk the same fine line as aristocrats, in that the right *type* of ill health must be maintained.

The novel ends, aptly, with the death of Godolphin who drowns after being struck on the back of the head while trying to cross a river: 'At daybreak, his corpse was found by the shallows of the ford; and the mark of violence across the temples, as of some blow, led them to guess that in scaling the banks, his head has struck against one of the tossing boughs that overhung them.'[88] This violent demise fully cements both Godolphin and Constance with their liminal class status—throughout the entire text, the threat of death hangs over every character associated with the aristocracy, and yet as Godolphin returns more to his middle-class roots, the threat manifests itself from without instead of within. His death fulfils the malign prophecy made by Volktman, a sort of cosmic revenge for Godolphin's ill romantic treatment of Volktman's low-ranking daughter, both mirroring the class-revenge premise that begins the novel and illustrating on what shaky medicalised grounds the aristocracy—and those who presume to associate with them—now stands.

Cheveley

Rosina Bulwer Lytton's 1839 novel, *Cheveley: or, The Man of Honour*, serves as an important counterpoint to Edward Bulwer-Lytton's *Godolphin*. Rosina's text, published six years after Edward's silver fork novel, is firstly an exemplar of later and better-established silver fork tropes and styles. It is secondly a direct and particularly scandalous rebuttal of *Godolphin* (in which the cold-hearted commoner, Constance, bore a striking resemblance to Rosina); to say these two novels were in dialogue with each other would be an understatement. Thirdly, her novel also undertakes a significant interrogation of the medicalisation of aristocrats, the sophistication and clarity of which matches the establishment of the genre and the further transition of intraclass mores and understandings of pathology.

Cheveley was written after years of unhappiness in her marriage. This included a notorious incident of domestic abuse, in which Rosina was convinced Edward meant to kill her; Edward's attempted reconciliation during a second honeymoon with Rosina was undermined by him bringing his mistress along, culminating in their separation in 1836.[89] Distressed by the very likely possibility of losing custody of her children, Rosina Bulwer Lytton's *Cheveley* was, out of necessity, one of the silver fork genre's most thinly veiled exposés and critiques of the aristocratic system, leading to her 'social ostracisation from the *bon ton*'.[90] The novel was written in an attempt to reveal Edward's cruelty to the reading public and to thereby influence any divorce proceedings by having their marriage tried more fully in the public sphere than an already highly public aristocratic divorce would have been—an amount of publicity fully exploited by Rosina's publisher who hoped that 'if it was opposed or prosecuted for libel, huge sales would be guaranteed'.[91] Rosina said of her estranged husband, 'Exposure is the only thing that complex monster dreads, and consequently the only check I have upon him.'[92] That Rosina's novel, which focuses on the sexist and classist elements of medicine, illness, and injury in high society, became even more entrenched in those very medical discourses through Edward's commitment of Rosina to an asylum due to her 'unfitness' as a mother and the supposed 'mental aberration' she possessed that led her to write such a novel.[93]

The novel's plot parallels that of *Godolphin* closely: the heroine, Lady Julia, begins life as a commoner and is brought into elite society through her advantageous marriage to a lord. Unlike Edward Bulwer-Lytton's Constance, however, Julia is not a destructive force to the aristocracy, nor does she even particularly wish to belong to that class; her abusive husband courts her relentlessly until she marries him, after which he beats and mistreats her, leading to her severe injury and illness. Just as in *Godolphin*, the marriage between the unworthy aristocrat and the plebeian beauty keeps her from a romance with the hero whose name gives title to the novel. In both novels, the aristocratic husband dies suddenly (in *Cheveley*'s case, suffering a particularly bloody fall from his horse), freeing his wife to remarry more judiciously. Unlike *Godolphin*, where the hero's life is ended due to the complex effects of class embodiment, *Cheveley* has a happy ending, further undoing much of Edward Bulwer-Lytton's discourses on class and health: in *Cheveley*, aristocrats only causes misery and ill heath to those more vulnerable than them, while revelling in the privilege of their own good health. Lady Julia, though affected by illness and feebleness during her aristocratic marriage, is able to remove the contagion of aristocracy and go on to have a pleasant and full life, while Edward Bulwer-Lytton's Constance forever bears the taint of crossing class boundaries. In both instances, class miscegenation leads to despair, with the aristocracy serving as a pathological locus, but the application and effects of this pathology differ greatly between texts.

Where Rosina was very likely an influence upon Edward's depiction of Constance, Lady Julia's villainous husband, Lord De Clifford, is modelled overtly on Edward; Rosina, in fact, seems to revel in antagonising him. She dedicates the novel to 'No One Nobody, Esq., of No Hall, Nowhere', directly needling Edward's pride and sensitivity about his family's exalted lineage.[94] Further in this vein, Rosina, 'kills off her husband [in the fictionalised guise of Lord De Clifford] in a scene where characters find themselves stepping in his spilt blood. The significance of this to Bulwer, who was proud of his noble blood, would not have gone unnoticed.'[95] Rosina further alienated her social circle by basing Lord De Clifford's lackeys on Edward's associates. Writing of one of her husband's sycophantic friends, thinly veiled in the novel as the character 'Major Nonplus', Rosina described his habit of 'taking his *appetitenal* [*sic*] walk before dinner, and looking, in his red Belcher cravat, Flamingo [*sic*] face, and scarlet waistcoat, for all the world like an ambulating carbuncle, trying to extinguish the setting sun'.[96] To Bulwer Lytton, Major Nonplus's regular pursuit of ruddy-cheeked good health not only might serve to extinguish that health, but also eradicates any sense of elegance or fashion. By aligning him with a carbuncle, Rosina Bulwer Lytton illustrates a metric by which health and illness might be assessed (the vulgarity of Nonplus's hearty activity and carbuncle complexion, versus the later swooning delicacy of the novel's hero and heroine), as well as how good health and robust constitutions work against both life and social livelihood.

More scathingly in the novel, Rosina Bulwer Lytton satirises her husband's family (particularly Edward's mother) through the character Lady De Clifford, the heroine's own dreadful mother-in-law. Just as with Rosina Bulwer Lytton's own strained familial relationships, it is not just the mere presence, but the strapping longevity of the De Clifford family that has a deleterious effect on the health and happiness of those who marry into that family. Bulwer Lytton writes of her long-lived characters that Lady De Clifford's father refused to die and thereby failed to leave Lady De Clifford and her new husband with wealth and status. When her father finally died, Lady De Clifford's husband barely 'survived that even six years, [and] quitted the world […], as some hinted, from the daily dose of this ungilded pill'.[97] Medicine and health are clearly framed in the light of villainy here, with the ill-tempered Lady de Clifford and her tenaciously vital father stopping an overdue inheritance and driving her husband into the grave, with Lady de Clifford paradoxically described as a curative (albeit an abrasive one). Again, health seems, counterintuitively, to come at the cost of both life and livelihood. The senior Lady De Clifford is later described as wearing 'a bottle-green cloth habit, studded with flat green cloth ipecacuana lozenge [a purgative remedy] looking buttons', muddying the waters between her metaphorical status as both an illness and its cure.[98]

Both Edward's and Rosina's novels indicate that illness is somehow a social contract or the byproduct of interaction: what illnesses characters contract or spread, and how ill people are treated, communicate to the reader everything they need to know about characters' relationships and socio-economic status. Further, both texts root illness in the liminal space where the middle classes and the aristocracy begin to overlap and where changing mores and practices become embodied. Where Edward ties illness firmly to the aristocracy (and hints that, though a mixed blessing, illness is something his middle-class characters are lucky to contract), Rosina takes a more middle-class—if subversively old-fashioned—perspective. Illness, to Rosina, is the purview of those who are noble in spirit even if they are not noble in lineage; it so follows that all of her aristocratic characters are overly healthy brutes, tying fashionable illness more to the realm of the upwardly ascendant middle classes. Initially her adherence to concepts of sensibility—exemplified by the sensitive Cheveley and Julia—read as a middle-class author attempting to espouse an outdated mode of aristocratic behaviour through a lack of proper class knowledge and that she mimics the medicalised ideals of a previous generation. Despite Rosina's antiquated representation of classed medicine, she uses the cult of sensibility to comment on changing socio-economic factors, especially as they concern masculinity. Marie Mulvey-Roberts writes in her introduction to the novel:

> The silver fork novel was in the process of defining the gentleman, an evolution that began with the Regency dance and culminated in the Victorian patriarch. Gentility was increasingly seen as depending upon meritocratic models of behaviour, rather than on blood and breeding. [Edward] Bulwer stood at the cross-roads of these differing constructions of masculinity.[99]

Rosina Bulwer Lytton, although unable to fully do away with illness's connection to the aristocracy (the middle-class Julia marries a lord, and the middle-class Cheveley eventually becomes a Marquis), does pair it with those who are upwardly mobile. More significantly, she denies the same form of glamorised illness to her nobly born characters, indicating that they've lost the sense of their class origins. Rather than the middle classes mimicking the embodied behaviour of their betters, illness has been taken, earned, or been ceded from one class group to another ascendant, more deserving one. To Rosina, Cheveley and Julia are reified as a new aristocracy because they exhibit a positive blend of old-fashioned, sentimental traits with modern conceptions of gentility, and miss out the lacklustre, contemporaneous aristocratic modes.

This is not to say that illness is never experienced by other class groups in Bulwer Lytton's text. Mademoiselle D'Antonville, governess to Julia and Lord De Clifford's daughter, is frequently ill, but in a way notably different from the refined and sympathetic Julia. Mademoiselle D'Antonville, who is transparently having an affair with Lord De Clifford, uses her bouts of headache and fainting (whether genuine or affected) to continue their flirtation, even in front of his wife. When the three of them take an uncomfortable carriage journey together,

> They had not got above half way, before mademoiselle began to purse up her mouth, close her drab-coloured eyes, and incline her head faintingly towards his shoulder, at which Lady [Julia] De Clifford offered her 'vinaigrette,' intending to request that she would change places with her, as she feared that sitting with her back to the horses might have occasioned her indisposition [...] Meanwhile mademoiselle, after labouring for a few minutes like a steam-engine, thought fit to open her eyes, and raising her head from Lord De Clifford's shoulder, where it had *unconsciously* rested, murmured, or rather shrieked.[100]

The crassness of Mademoiselle D'Antonville's behaviour surrounding her illness, if not necessarily the illness itself, bespeaks her poor breeding and aligns her to the equally boorish aristocrats who are either never ill, bring their illnesses on themselves in the form of hangovers and gout, or handle their illnesses as coarsely as they treat the protagonists of the novel. Julia, meanwhile, illustrates her own refinement many times over in this passage. Materially, she has a vinaigrette close to hand, and a knowledge of some situational causes of the indisposition, bespeaking her own previous experience with fainting. In terms of delicacy of feeling, Julia's 'intention' to request that Mademoiselle D'Antonville swap places with her implies a lack of guile in Julia and an inability to see it in anyone else. Although Julia could hardly be blamed for requesting the swap due to Mademoiselle D'Antonville's impropriety in laying her head on Lord De Clifford's shoulder, the idea seems not to have occurred to Julia, who has a genuine concern for the governess's well-being.

The most famous episode of the novel divides sensibility from the aristocracy most fully: during a domestic disagreement, Lord De Clifford becomes violent and strikes Julia's hand, spraining her wrist. The subsequent shock to the gentle Julia's system nearly kills her and, by extension, nearly kills the empathetic Cheveley, who rushes to find her medical assistance:

> [T]ottering toward a sofa, she sank fainting and exhausted upon it, the mask falling at the same time from her lifeless hand [...] There lay before him, helpless and unconscious, all that he loved on earth, [...] And raising the loose sleeve of her domino, her swollen, drooping, and blackening right hand met his view.[101]

After a frantic search for a surgeon, who initially worries about Cheveley's own health due to his pale, worried countenance, they apply leeches to her hand, give her draughts, and force her to remain in bed for weeks; despite this careful attention, her recovery is slow. Although it is now ridiculous to contemplate a physiology so delicate that a sprained wrist could place one at the brink of death (and, indeed, this episode is a more melodramatic and yet less physically grave reworking of Rosina's own serious domestic abuse at the hands of Edward), it is in alignment with the class and body parameters set out in the rest of the novel. Julia and Cheveley operate within their own small, medically and morally sensitive group, where class and generational mores overlap to create a new elite.

Cecil

Coming at the tail end of the silver fork genre, Catherine Gore's *Cecil, or the Adventures of a Coxcomb* (1841) satirises what April Kendra defines as the 'dandy novel': a subset of silver fork fiction which focuses less on the marriage plot formula and more on the vaguely picaresque escapades of regency bucks.[102] Of all the silver fork novels and their parodies, *Cecil* is perhaps the most overt in its commentary on aristocratic pathology, with significant sections of the novel centring around medical farce. This is perhaps unsurprising, as Gore—who was easily one of the genre's most famous, prolific, and critically and commercially successful writers—had a keen awareness of the tropes of the genre she helped to build with her early successes, and helped to undermine with her later satires.[103]

The eponymous Cecil, the snobbish and profligate younger brother of a lord, spends a majority of the novel fickly pursuing love affairs and coping with their dissolution through varying degrees of self-destructiveness or convalescence. While Cecil's reactions, both behavioural and physiological, are often comically melodramatic given the shallowness of his romantic attachments, his escapades are frequently punctuated by death. Three of his conquests die—albeit of ambiguous causes—and Cecil is the direct, if accidental, catalyst for his young nephew's fatal fall from a horse; Cecil ends the narrative a confirmed bachelor and the heir presumptive to his brother's estate. Gore's portrait in *Cecil* is of an aristocracy that survives instead of thrives, and remains deeply confused about its own relationship to the medical and fashionable landscape even as it asserts several competing theories about the nature of that relationship.

Cecil departs from the more traditional portrayals of medicalised class in silver fork novels in that its aristocrats argue—tongue-in-cheek on the part of Gore—that the true sign of an aristocrat was robust good health and physical perfection. Early in the novel, Cecil says of his siblings that the 'Honourable John squinted, and the Honourable Julia had red hair; and our lady-mother was as heartily ashamed of them both, as if they had been palmed upon her from the workhouse'.[104] His elder brother's poor eyesight and his sister's homeliness, compared to Cecil's own physical excellence, initially indicate that he should be the true heir to the title and is wasted in the role of the 'spare'. Despite Cecil's egotism on this point, he is seemingly aware of—and yet untouched by—the subtle implication that he might be illegitimate and the product of his mother's affair with a non-aristocratic admirer. Cecil, with the socio-economic backing to deny any potential middle-class lineage in himself, carries on with his thesis. Any sign of physical delicacy is merely social-climbing performativity (if fake) or a lack of good blood and breeding (if genuine). Cecil states, 'I am of the opinion that a sea-sick man was not born to sail in his own yacht; and am proud to declare that the heaviest swell finds me enjoying the robust health becoming a gentleman.'[105] Far from entertaining the notion that sensibility might be a genuine barometer of aristocracy, he instead couples health with both consumer goods and the leisure time needed to enjoy them. Again Cecil's argument is undone, many times over: though said with an air of authority, Cecil states that *he* is of the *opinion*, not that his hale-and-hearty physiology is a widely accepted marker of aristocracy. Further, he fails to acknowledge that he does, in fact, get ill at various points in the novel and that these illness are sometimes the result of emotional strain and therefore sentimental in nature. Finally, his potential illegitimacy makes no mark on his views of his own lineage and bodily health (and is perhaps wilfully ignored on his part); what is particularly obfuscating in Cecil's decree is the vague manner in which he describes himself as a 'gentleman'. There was, as has been previously discussed, a paradigm shift in the first half of the nineteenth century when it came to the concept of a gentleman, so it is difficult to parse whether Cecil—equal parts hyper-contemporary and pompous—meant 'gentleman' in its more modern context as an appellation accessible to the middle classes, or if he meant it in its more traditional, upper-class sense. As the novel is told in hindsight, it's even more difficult to understand if Cecil is merely recounting what he thought at the time, years previously, or if it is what he as a narrator currently believes.

Despite Cecil's insistence that his aristocratic pedigree keeps him healthy, he repeatedly aligns himself with death. Opening the novel, Cecil warns the reader of the critics: 'Ten to one, they [the critics] will try to Burke my book, and render it the prey of the resurrectionists [...] dissecting me to ornament their hideous museum.'[106] That he begins at his own theoretical ending before he even recounts his own birth, with the pages of his memoirs ripe for a metaphorical autopsy table, makes the reader all the more aware of the cyclical and retrogressive nature of both the silver fork novels and the aristocracy as an apparatus: it requires constant turnover to maintain it, but there is little or no progress. Cecil is even, at one point in the novel, reported dead to his family after being caught up in a battle in Spain during one of his long, self-indulgent trips to the Continent. He returns home to find his family in mourning, and his mother's delight at his sudden appearance has nothing to do with the salvation Cecil's life or the future of their family, but rather because she is no longer obliged to wear unfashionable mourning attire.[107] Gore, commenting on the long-term viability of the aristocracy as an institution, articulates that aristocratic death is rooted in the present, with only the immediacy of socio-economic benefits or disadvantages being of any importance to its members.

Crucially, it is not Cecil who is the most affected by illness and death, although he does at times seem to be at the epicentre of all medical incidents in the novel. Those most affected are either those more clearly aligned with the upper classes (his nephew and heir to the family title and estate, who dies) or those more clearly aligned with the middle or lower classes (his three main love interests, Emily, Franszetta, and Helena, all of whom die). Regarding the former, there is a certain traditional upper-class frailty associated with the boy, especially since his more robust and potentially more middle-class uncle survives far worse over the course of the novel than a fall from a horse. However, aristocracy is here confused, as Cecil's older brother, though the legitimate holder of the title, does not participate in aristocratic, or at least stylish, society in a manner Cecil thinks befitting his station. Here ideas of lineage and the body become confused as to which of the two brothers is more contaminated by class mobility or class ambiguity. Regarding Cecil's unambiguously middle- and lower-class love interests, Gore portrays a greater sense of pathological dialogue: it is their association with Cecil that appears to be the catalyst for their unspecified, terminal, sensibility-coded ailments, and their deaths lead to Cecil's own physical breakdowns and illnesses. A new romantic partner serves as a temporary cure, at which point the cycle repeats itself. Gore's middle- and lower-class women are ultimately disposable and replaceable. Cecil, who gloats early in the novel, 'I thank Heaven, I was born a coxcomb, for coxcombs are bachelors by prescriptive right', has no intention of forging a marriage, family, or even a genuine human connection with any woman.[108] He instead creates impotent, cyclical melodrama, using middle- and lower-class women to provide him with distractions. Gore mimics Cecil's utilisation of these characters: the novel requires their presence and deaths to serve as plot devices in a narrative that ultimately, intentionally, goes nowhere. Cecil, like the aristocracy he seeks to embody, learns little and changes even less.

More than the protagonist's embodiment of the aristocratic institution—which, through Cecil, is depicted as merely surviving but having no real future—Gore's *Cecil* also epitomises the stagnating silver fork genre itself which would, indeed, spin its wheels for another decade before fading out. It is through Cecil's understanding of class-based pathology (or, rather, its comfort in depicting a *lack* of understanding) that Gore most effectively skewers the upper classes and silver fork romances. *Cecil* is full of vacillation, and its arguments on class-based medicalisation are deliberately muddy and contradictory. Gore depicts both a class group and a genre in which there no longer any consistent connection between social status and health, even in terms of performativity. The genre, which contributed to social flux and the confusion of class codifiers, can no longer stand under the weight of its own success.

Conclusion

The silver fork novels, far from their implied objective to clarify upper-class bodies and behaviour, more often complicated their subjects than not. At times, this is because the novels outright mislead readers (through some of the novels' tongue-in-cheek humour, through the protection of genuine class practices by aristocratic authors, or by providing plausible-sounding guesswork advice from authors whose only goal was to push sales figures). Much of the misinformation, however, was the result of the novels not being able to live up to the promises set out by their advertisers: the practices, opinions, and understandings of bodies and health were not often cohesive and clear, any more than those who qualified as 'the upper ten thousand' over several decades were a single homogenous group.

What is significant and consistent when looking at this genre is that aristocratic bodies were considered very frequently (possibly to the extent of *every* instance within the genre) to be worthy of medical scrutiny in a way that their middle- and lower-class character counterparts were not. It is perhaps unsurprising that as class boundaries began to fracture, the 'reality' of embodied aristocracy became a greater concern, with class signifiers scrutinised for authenticity and sought for the social clarity they could provide. The silver fork genre tracks both the overtly discussed and the deeply internalised shifting attitudes about a pathological aristocracy, from sentimental frailty to robust superiority, from one fashionable medical issue to another, from performance to embodiment, and back again. As we will see in subsequent chapters, the vagaries produced by this genre did not result in an eschewal of the topic, but rather a reification and reclassification. Aristocratic bodies and health serve as socio-economic signifiers (or at least are *treated* as keys to decode socio-economic ciphers), although the particular concerns the public applies to these bodies are as unstable and oscillating as the readings and realities of the bodies themselves.

Notes

1. William Makepeace Thackeray, *The Book of Snobs* (1846–47), *The Works of William Makepeace Thackeray*, Vol. 15 of 22 (London: Smith, Elder & Co, 1869), p. 133.
2. William Makepeace Thackeray, *Vanity Fair* (1848), ed. by John Carey (New York: Penguin, 2004), p. 216, emphasis original.
3. Clare Bainbridge, 'Introduction' to *Granby* by Thomas Henry Lister (1826) (London: Pickering & Chatto, 2005), pp. xxiii–xxxv (p. xxii).

4. Cheryl A. Wilson, *Fashioning the Silver Fork Novel* (London, Pickering & Chatto, 2012), p. 1.
5. Susan Sontag, *Illness as Metaphor* (1978) (London: Penguin, 2002), p. 3.
6. Sumiao Li, '"Arabian Nights Entertainment": The Rule of Fashion and the Public Roles of Aristocratic Women in Britain 1820–1860', *Nineteenth-Century Gender Studies* 6:1 (Spring 2010), available at: http://www.ncgsjournal.com/issue61/li.htm [accessed 10 September 2019].
7. Cynthia Lawford, 'Introduction' to *Romance and Reality* by Letitia Landon (1831) (London: Pickering & Chatto, 2005), pp. ix–xxvii (p. xvii); Marie Mulvey-Roberts, 'Introduction' to *Cheveley; or, The Man of Honour* by Rosina Bulwer Lytton (1839) (London: Pickering & Chatto, 2005), pp. ix–xxvii (pp. xxiv–xxv); Edward Copeland, *The Silver Fork Novel: Fashionable Fiction in the Age of Reform* (Cambridge: Cambridge University Press, 2012), p. 19; Tamara S. Wagner, *Financial Speculation in Victorian Fiction* (Columbus: Ohio State University Press, 2010), p. 32.
8. Winifred Hughes, *'Silver Fork Writers and Readers: Social Contexts of a Best Seller'*, *NOVEL: A Forum on Fiction* 25:3 (Spring 1992), pp. 328–47 (p. 331).
9. 'Aristocracy', *The New Monthly Magazine and Literary Journal* 35:140 (August 1832), pp. 163–67 (pp. 163–65, emphasis original).
10. Tamara S. Wagner, 'Silver-Fork Legacies: Sensationalizing Fashionable Fiction', *Women's Writing* 16:2 (2009), pp. 301–22 (p. 302).
11. Patrick Brantlinger, *The Spirit of Reform: British Literature and Politics, 1832–1867* (Cambridge: Harvard University Press, 1977), p. 120.
12. Marrisa Joseph, *Victorian Literary Businesses: The Management and Practices of the British Publishing Industry* (Palgrave Macmillan: London, 2019), p. 21.
13. Thorstein Veblen, *The Theory of the Leisure Class: An Economic Study in the Evolution of Institutions* (1899). (Waiheke Island: Floating Press, 2009), pp. 83–121.
14. Bainbridge, pp. xxxii–iii.
15. Copeland, p. 6.
16. Abigail Boucher, "The Business Model of the Aristocracy: Class, Consumerism, and Commodification in the Silver Fork Novels", *Nineteenth-Century Contexts* 38:3 (2016), 1–10.
17. "Aristocratic Revelations", *Westminster Review* 22:44 (April 1835), pp. 314–21 (p. 314).
18. Matthew Whiting Rosa, *The Silver-Fork School: Novels of Fashion Preceding* Vanity Fair (1936) (Port Washington, New York: Kennikat Press, 1964), p. 7.
19. See Copeland, *The Silver Fork Novel;* Alison Adburgham, *Silver Fork Society: Fashionable Life and Literature from 1814 to 1840* (1983)

(London: Faber and Faber, 2012); the *Silver Fork Novels, 1826–1841* series, ed. by Harriet Devine Jump, 6 vols (London: Pickering & Chatto, 2005); Rosa's *The Silver-Fork School: Novels of Fashion Preceding* Vanity Fair (1936); April Kendra's, 'Gendering the Silver Fork: Catherine Gore and the Society Novel', *Women's Writing*, 11:1 (2004), pp. 25–38 and "Silver-Forks and Double Standards: Gore, Thackeray and the Problem of Parody', *Women's Writing*, 16:2 (2009), pp. 191–217 (p. 191);. Wilson, *Fashioning*; and Hughes's 'Silver Fork Writers and Readers'.

20. William Hazlitt, 'The Dandy School', *The Examiner*, 1033 (18 November, 1827), pp. 721–23 (p. 722).
21. Charlotte Brontë, *Jane Eyre* (1848), ed. by Beth Newman (New York: Bedford/St. Martin, 1996), p. 234; Thomas Carlyle, *Sartor Resartus* (1834), ed. by Kerry McSweeny and Peter Sabor (Oxford: Oxford University Press, 2008), pp. 210–11.
22. Wilson, *Fashioning* (London: Pickering & Chatto, 2012), p. 20.
23. Sontag, *Illness*, p. 28.
24. Miriam Bailin, *The Sickroom in Victorian Fiction: The Art of Being Ill* (Cambridge: CUP, 1994), p. 8.
25. George C. Grinnell, *The Age of Hypochondria: Interpreting Romantic Health and Illness* (Houndmills, Basingstoke: Palgrave Macmillan, 2010), p. 5; Bailin, *Sickroom*, p. 8; Andrew Scull, *Hysteria: The Biography* (Oxford: Oxford University Press, 2009), pp. 51–52; Michael Stolberg, *Experiencing Illness and the Sick Body in Early Modern Europe* (Houndmills, Basingstoke: Palgrave Macmillan, 2011), pp. 181–84; Wayne Wild, *Medicine-by-Post: The Changing Voice of Illness in Eighteenth-Century British Consultation Letters and Literature* (New York: Rodopi, 2006), pp. 10–12.
26. Stolberg, *Experiencing Illness*, pp. 181–82.
27. Michel de Montaigne, 'Not to Counterfeit Being Sick', *Essays of Michel De Montaigne* (1592), Trans. Charles Cotton, ed. William Carew Hazlitt (1877), (n.p., Book II, Chapter XXIV).
28. Roy Porter, *Madness: A Brief History* (Oxford: Oxford University Press, 2002) p. 86.
29. George Cheyne, *The English Malady: or, as a Treatise of Nervous Diseases of all* Kinds (London: G. Strahan, 1733), pp. 158–59.
30. Grinnell, *Hypochondria*, p. 37; Thomas Beddoes, *Hygëia: or Essays Moral and Medical on the Causes Affecting the Personal State* (Bristol: R. Phillips, 1802).
31. Athena Vrettos, *Somatic Fictions: Imagining Illness in Victorian Culture* (Stanford, CA: Stanford University Press, 1995), p. 2.
32. M.L. Bush, *Rich Noble, Poor Noble* (New York; Manchester University Press, 1988), p. 5; Andrew Adonis, *Making Aristocracy Work* (Oxford:

Clarendon Press, 1993), p. 1; David Cannadine, *Aspects of Aristocracy: Grandeur and Decline in Modern Britain* (London: Yale University Press, 1999), p. 2.
33. Katherine Byrne, *Tuberculosis and the Victorian Literary Imagination* (Cambridge: Cambridge University Press, 2011), p. 94.
34. Thomas Rowlandson, 'The Genealogist', *The English Dance of Death* (1814–16) (London: Methuen and Co, 1903), pp. 118–26 (pp. 118–19).
35. A.N. Wilson, *The Victorians* (London: Arrow Books, 2002), pp. 10–11; Andrew Adonis, *Making Aristocracy Work* (Oxford: Clarendon Press, 1993), p. 240; Toke S. Aidt and Raphaël Franck, "Democratization under the Threat of Revolution: Evidence from the Great Reform Act of 1832", *Econometrica* 83:2 (March 2015), pp. 505–47 (p. 509).
36. Marguerite Gardiner, The Countess of Blessington, *The Confessions of an Elderly Gentleman* (London: Longman, Rees, Orme, Brown, Green, and Longman, 1836), p. 3; Charlotte Trimmer Moore, *Country Houses*, vol. 3 of 3 (London: Saunders and Otley, 1832), p. 29, emphasis original.
37. Roy Porter and George Sebastian Rousseau, *Gout: The Patrician Malady* (London: Yale University Press, 1998), p. 56.
38. Ibid., p. 57; p. 72.
39. C.D. Burdett, *At Home*, vol. 1 of 3 (London: Henry Colburn, 1828), pp. 41–42.
40. Tamara S. Wagner, '"Anything but business": Middle-Class Work and Leisure in Catherine Gore's Silver-Fork Fiction', *Victorians: A Journal of Culture and Literature* 137 (Summer 2020), pp. 1–14.
41. Michel Foucault, *The History of Sexuality* (1976), trans. by Robert Hurley, vol. 1 of 4 (New York: Pantheon Books, 1978) p. 124 (emphasis original).
42. Roy Porter, *Health for Sale: Quackery in England 1660–1850* (Manchester: Manchester University Press, 1989), p. 41; Grinnell, *Hypochondria*, p. 18.
43. Grinnell, *Hypochondria*, p. 8.
44. Catherine Gore, *Cecil, or the Adventures of a Coxcomb* (1841), ed. by Andrea Hibbard and Edward Copeland (London: Pickering & Chatto, 2005), pp. 289–90.
45. Letitia Landon, *Romance and Reality* (1831), ed. by Cynthia Lawford (London: Pickering & Chatto, 2005), p. 126.
46. Copeland, *Silver Fork*, p. 138.
47. Anonymous, *High Life*, vol. 1 of 3 (London: Saunders and Otley, 1827), p. 4.
48. Trimmer Moor, *Country Houses*, vol. 1 of 3 (London: Saunders and Otley, 1832), pp. 39–40.
49. Joseph, *Victorian Literary Businesses*, p. 30.
50. Ibid., pp. 30–31; p. 54.

51. Cheryl A. Wilson, *Fashioning*, pp. 80–81.
52. Porter, *Health*, p. 111.
53. Porter, *Health*, p. 91.
54. Rowlandson, 'The Quack Doctor', *The English Dance of Death* (1814–16) (London: Methuen and Co, 1903) pp. 74–83 (p. 77).
55. Porter, *Health*, p. 19.
56. Genice Ngg, 'The Changing Face of Quack Doctors: Satirizing Mountebanks and Physicians in Seventeenth- and Eighteenth-Century England', *New Directions in Literature and Medicine Studies*, ed. Stephanie M. Hilger (London: Palgrave Macmillan, 2017), pp. 333–56 (p. 340).
57. Charles Dickens, *A Tale of Two Cities* (1859) (Ware: Wordsworth Classics, 1993), p. 89.
58. Bailin, p. 9
59. Stolberg, *Experiencing Illness*, p. 63; Wild, *Medicine-by-Post*, p. 19.
60. Porter, *Health*, p. 47.
61. Harriet Devine Jump, "General Introduction" to *The Silver Fork Novels, 1826–1841* in *Granby* (1826) by Thomas Henry Lister, ed. by Clare Bainbridge, vol. 1 of 6 (London: Pickering & Chatto, 2005), p. ix; Andrea Hibbard and Edward Copeland, "Introduction" to *Cecil* by Catherine Gore (1841), ed. by Andrea Hibbard and Edward Copeland (London: Pickering & Chatto, 2005), p. X; Lawford, "Introduction", p. xiii; Copeland, *Silver Fork*, p. 16; Wagner, *Financial Speculation*, p. 32.
62. Lawford, 'Introduction', p. xviii.
63. Copeland, *Silver Fork*, p. 24.
64. Thomas Percival, *Medical Ethics: Or, a Code of Institutes and Precepts, Adapted to the Professional Conduct of Physicians and Surgeons* (Cambridge: Cambridge University Press, 1803), pp. 44–45.
65. Anonymous, 'Compositions of Quack Medicines', *The Lancet* 1:4 (26 October 1823), p. 138.
66. Porter, *Health* (p. 16; Sylvia A. Pamboukian, *Doctoring the Novel: Medicine and Quackery from Shelley to Doyle* (Athens OH: Ohio University Pres, 2012), pp. 9–10.
67. Trimmer Moore, *Country Houses*, vol. 3 of 3, p. 29.
68. Lady Charlotte Campbell Bury, *The Exclusives*, vol. 2 of 3 (London: Henry Colburn and Richard Bentley, 1830), p. 2.
69. 'Somatic Symptom and Related Disorders', *The Diagnostic and Statistical Manual of Mental Disorders, Fifth Edition* (Washington, D.C.: American Psychiatric Association, 2013), pp. 310–27 (p. 311).
70. The father of the titular hero, Godolphin, is described as 'dressed in an old flannel dressing-gown […] and cowering gloomily over a wretch fire, [a] mixture of half hypochondriac, half miser, which he was in reality'

(p. 13). Later in the novel, a fashionable society dame is described as 'always employed in nursing her own health: hypochondria supplies her with an object; she is really happy, because she fancies herself ill' (p. 161).

71. Elaine Scarry, *The Body in Pain: The Making and Unmaking of the World* (Oxford: OUP, 1985), p. 13.
72. Jump, "Introduction", p. xv.
73. Leslie Mitchell, *Bulwer Lytton: The Rise and Fall of a Victorian Man of Letters* (London: Hambledon and London, 2003), p. 31; Alison Adburgham, *Silver Fork Society*, p. 135.
74. Sir Edward Bulwer-Lytton, 'Preface' to *Godolphin* by Sir Edward Bulwer-Lytton (1833), (Boston: Little, Brown, and Company, 1899) pp. xiii–xvii (p. xiv).
75. Sir Edward Bulwer-Lytton, p. 6.
76. Ibid., p. 7.
77. Ibid., p. 6.
78. Ibid., p. 5.
79. Ibid., p. 5.
80. Ibid., p. 6.
81. Ibid., p. 8.
82. Ibid., p. 134.
83. Ibid., p. 180.
84. Ibid 215.
85. Ibid., p. 24; p. 31, pp. 247–48.
86. Ibid., p. 136.
87. Ibid., p. 284.
88. Ibid., p. 312.
89. Mulvey-Roberts, pp. xviii–xx.
90. Ibid., p. xvii.
91. Ibid., xiii.
92. Ibid., xii.
93. Ibid., xvii.
94. Rosina Bulwer Lytton, *Cheveley; or, The Man of Honour* (1839) (London: Pickering & Chatto, 2005)p. 3.
95. Mulvey-Roberts, pp. xx.
96. Rosina Bulwer Lytton, *Cheveley*, p. 15.
97. Ibid., p. 70.
98. Ibid., p. 107.
99. Mulvey-Roberts, pp. xxiv.
100. Ibid., pp. 57–58.
101. Rosina Bulwer Lytton, *Cheveley*, pp. 214–15.
102. Kendra, 'Gendering', p. 25.

103. Molly Engelhardt, "'The Novelist of a New Era': Deepening the Sketch of Catherine Gore", *Victorian Review* 42:1 (Spring 2016), pp. 65–84 (p. 65); Catherine Gore wrote at least one, and as many as five novels, per year from 1829 until 1847, after which her writing speed drastically slowed.
104. Gore, *Cecil*, p. 8.
105. Ibid., p. 154.
106. Ibid., p. 4.
107. Ibid., p. 158.
108. Ibid., *Cecil*, p. 14.

Works Cited

Adburgham, Alison, *Silver Fork Society: Fashionable Life and Literature from 1814 to 1840* (1983) (London: Faber and Faber, 2012).

Adonis, Andrew, *Making Aristocracy Work* (Oxford: Clarendon Press, 1993).

Aidt, Toke S., and Raphaël Franck, 'Democratization under the Threat of Revolution: Evidence from the Great Reform Act of 1832', *Econometrica* 83:2 (March 2015), pp. 505–47.

Anonymous, 'Compositions of Quack Medicines', *The Lancet* 1:4 (26 October 1823), p. 138.

Anonymous, *High Life*, 3 vols. (London: Saunders and Otley, 1827).

'Aristocracy', *The New Monthly Magazine and Literary Journal* 35:140 (August 1832), pp. 163–67.

'Aristocratic Revelations', *Westminster Review* 22:44 (April 1835), pp. 314–21.

Miriam Bailin, *The Sickroom in Victorian Fiction: The Art of Being Ill* (Cambridge: CUP, 1994).

Bainbridge, Clare, 'Introduction' to *Granby* by Thomas Henry Lister (1826) (London: Pickering & Chatto, 2005), pp. xxiii–xxxv.

Beddoes, Thomas, *Hygëia: or Essays Moral and Medical on the Causes Affecting the Personal State* (Bristol: R. Phillips, 1802).

Boucher, Abigail, 'The Business Model of the Aristocracy: Class, Consumerism, and Commodification in the Silver Fork Novels', *Nineteenth-Century Contexts* 38:3 (2016), 1–10.

Brantlinger, Patrick *The Spirit of Reform: British Literature and Politics, 1832–1867* (Cambridge: Harvard University Press, 1977).

Brontë, Charlotte, *Jane Eyre* (1848), ed. by Beth Newman (New York: Bedford/ St. Martin, 1996).

Bulwer-Lytton, Sir Edward, 'Preface' to *Godolphin* by Sir Edward Bulwer-Lytton (1833), (Boston: Little, Brown, and Company, 1899) pp. xiii–xvii.

Bulwer Lytton, Rosina, *Cheveley; or, The Man of Honour* (1839) (London: Pickering & Chatto, 2005).

Burdett, C.D., *At Home*, 3 vols. (London: Henry Colburn, 1828).

Bush, M.L., *Rich Noble, Poor Noble* (New York; Manchester University Press, 1988).

Byrne, Katherine, *Tuberculosis and the Victorian Literary Imagination* (Cambridge: Cambridge University Press, 2011).

Campbell Bury, Lady Charlotte, *The Exclusives*, 3 vols. (London: Henry Colburn and Richard Bentley, 1830).

Cannadine, David, *Aspects of Aristocracy: Grandeur and Decline in Modern Britain* (London: Yale University Press, 1999).

Carlyle, Thomas, *Sartor Resartus* (1834), ed. by Kerry McSweeny and Peter Sabor (Oxford: Oxford University Press, 2008).

Cheyne, George, *The English Malady: or, as a Treatise of Nervous Diseases of all* Kinds (London: G. Strahan, 1733).

Copeland, Edward, *The Silver Fork Novel: Fashionable Fiction in the Age of Reform* (Cambridge: Cambridge University Press, 2012).

Devine Jump, Harriet, 'General Introduction' to *The Silver Fork Novels, 1826–1841* in *Granby* (1826) by Thomas Henry Lister, ed. by Clare Bainbridge, vol. 1 of 6 (London: Pickering & Chatto, 2005), pp. 321–45.

Dickens, Charles, *A Tale of Two Cities* (1859) (Ware: Wordsworth Classics, 1993).

Engelhardt, Molly, '"The Novelist of a New Era": Deepening the Sketch of Catherine Gore' *Victorian Review* 42:1 (Spring 2016), pp. 65–84.

Foucault, Michel, *The History of Sexuality* (1976), trans. by Robert Hurley, 4 vols. (New York: Pantheon Books, 1978).

Gardiner, Marguerite, The Countess of Blessington, *The Confessions of an Elderly Gentleman* (London: Longman, Rees, Orme, Brown, Green, and Longman, 1836).

Gore, Catherine, *Cecil, or the Adventures of a Coxcomb* (1841), ed. by Andrea Hibbard and Edward Copeland (London: Pickering & Chatto, 2005).

Grinnell, George C. *The Age of Hypochondria: Interpreting Romantic Health and Illness* (Houndmills, Basingstoke: Palgrave Macmillan, 2010).

Hazlitt, William, 'The Dandy School', *The Examiner*, 1033 (18 November, 1827), pp. 721–23.

Hibbard, Andrea, and Edward Copeland, 'Introduction' to *Cecil* by Catherine Gore (1841), ed. by Andrea Hibbard and Edward Copeland (London: Pickering & Chatto, 2005), pp. ix–xxv.

Hughes, Winifred, *'Silver Fork Writers and Readers: Social Contexts of a Best Seller'*, *NOVEL: A Forum on Fiction* 25:3 (Spring 1992), pp. 328–47.

Joseph, Marrisa, *Victorian Literary Businesses: The Management and Practices of the British Publishing Industry* (Palgrave Macmillan: London, 2019).

Kendra, April, 'Gendering the Silver Fork: Catherine Gore and the Society Novel', *Women's Writing*, 11:1 (2004), pp. 25–38.

Kendra, April. 'Silver-Forks and Double Standards: Gore, Thackeray and the Problem of Parody', *Women's Writing*, 16:2 (2009), pp. 191–217.

Landon, Letitia, *Romance and Reality* (1831), ed. by Cynthia Lawford (London: Pickering & Chatto, 2005).

Lawford, Cynthia, 'Introduction' to *Romance and Reality* by Letitia Landon (1831) (London: Pickering & Chatto, 2005), pp. ix–xxvii.

Li, Sumiao, '"Arabian Nights Entertainment": The Rule of Fashion and the Public Roles of Aristocratic Women in Britain 1820–1860', *Nineteenth-Century Gender Studies* 6:1 (Spring 2010), available at: http://www.ncgsjournal.com/issue61/li.htm [accessed 10 September 2019].

Mitchell, Leslie, *Bulwer Lytton: The Rise and Fall of a Victorian Man of Letters* (London: Hambledon and London, 2003).

Montaigne, Michel de, 'Not to Counterfeit Being Sick', *Essays of Michel De Montaigne* (1592), Trans. Charles Cotton, ed. William Carew Hazlitt (1877), (n.p., Book II, Chapter XXIV).

Mulvey-Roberts, Marie, 'Introduction' to *Cheveley; or, The Man of Honour* by Rosina Bulwer Lytton (1839) (London: Pickering & Chatto, 2005), pp. ix–xxvii.

Ngg, Genice, 'The Changing Face of Quack Doctors: Satirizing Mountebanks and Physicians in Seventeenth- and Eighteenth-Century England', *New Directions in Literature and Medicine Studies*, ed. Stephanie M. Hilger (London: Palgrave Macmillan, 2017), pp. 333–56.

Pamboukian, Syliva A., *Doctoring the Novel: Medicine and Quackery from Shelley to Doyle* (Athens OH: Ohio University Pres, 2012).

Percival, Thomas, *Medical Ethics: Or, a Code of Institutes and Precepts, Adapted to the Professional Conduct of Physicians and Surgeons* (Cambridge: Cambridge University Press, 1803).

Porter, Roy, *Health for Sale: Quackery in England 1660–1850* (Manchester: Manchester University Press, 1989).

Porter, Roy. *Madness: A Brief History* (Oxford: Oxford University Press, 2002).

Porter, Roy, and George Sebastian Rousseau, *Gout: The Patrician Malady* (London: Yale University Press, 1998.

Rosa, Matthew Whiting, *The Silver-Fork School: Novels of Fashion Preceding* Vanity Fair (1936) (Port Washington, New York: Kennikat Press, 1964).

Rowlandson, Thomas, 'The Genealogist', *The English Dance of Death* (1814–16) (London: Methuen and Co, 1903a), pp. 118–26.

Rowlandson, Thomas. 'The Quack Doctor', *The English Dance of Death* (1814–16) (London: Methuen and Co, 1903b) pp. 74–83.

Scarry, Elaine, *The Body in Pain: The Making and Unmaking of the World* (Oxford: Oxford University Press, 1985).

Scull, Andrew, *Hysteria: The Biography* (Oxford: Oxford University Press, 2009).

'Somatic Symptom and Related Disorders', *The Diagnostic and Statistical Manual of Mental Disorders, Fifth Edition* (Washington, D.C.: American Psychiatric Association, 2013), pp. 310–27.

Sontag, Susan, *Illness as Metaphor* (1978) (London: Penguin, 2002).
Stolberg, Michael, *Experiencing Illness and the Sick Body in Early Modern Europe* (Houndmills, Basingstoke: Palgrave Macmillan, 2011).
Thackeray, William Makepeace, *The Book of Snobs* (1846–47), *The Works of William Makepeace Thackeray*, Vol. 15 of 22 (London: Smith, Elder & Co, 1869).
Thackeray, William Makepeace. *Vanity Fair* (1848), ed. by John Carey (New York: Penguin, 2004).
Trimmer Moore, Charlotte, *Country Houses*, 3 vols. (London: Saunders and Otley, 1832).
Veblen, Thorstein, *The Theory of the Leisure Class: An Economic Study in the Evolution of Institutions* (1899). (Waiheke Island: Floating Press, 2009).
Vrettos, Athena, *Somatic Fictions: Imagining Illness in Victorian Culture* (Stanford, CA: Stanford University Press, 1995).
Wagner, Tamara S., '"Anything but business": Middle-Class Work and Leisure in Catherine Gore's Silver-Fork Fiction', *Victorians: A Journal of Culture and Literature* 137 (Summer 2020), pp. 1–14.
Wagner, Tamara S. *Financial Speculation in Victorian Fiction* (Columbus: Ohio State University Press, 2010).
Wagner, Tamara S. 'Silver-Fork Legacies: Sensationalizing Fashionable Fiction', *Women's Writing* 16:2 (2009), pp. 301–22.
Wild, Wayne, *Medicine-by-Post: The Changing Voice of Illness in Eighteenth-Century British Consultation Letters and Literature* (New York: Rodopi, 2006).
Wilson, A.N., *The Victorians* (London: Arrow Books, 2002).
Wilson, Cheryl A., *Fashioning the Silver Fork Novel* (London, Pickering & Chatto, 2012).

CHAPTER 3

'Unblessed by Offspring': Fertility and the Aristocratic Male in Reynolds's *The Mysteries of the Court of London*

Introduction

The quotation which titles this chapter is found, in some form or another, in reference to nearly every aristocratic couple in the Chartist writer G.W.M Reynolds's 1840s–1850s serialised radical Chartist penny fiction, *The Mysteries of the Court of London*.[1] Silver fork fiction considered how the complex performances of aristocratic illness or functions of the body morphed through the generations as fashionable ideals evolved. Here, in Reynolds's working-class literature, those functions are given a much more definite, if no less complex, treatment. Reynolds still very much associates the aristocracy with ill health, but in the targeted (and ideologically charged) realm of fertility. In his biopolitical critique of the aristocratic system's ability to function, Reynolds's text frankly and bluntly places the blame for infertility upon the male partner, that it is 'the miserable husband [who] is impotent'.[2] This chapter analyses the manifestations of endemic aristocratic infertility in Reynolds's Chartist penny fiction and explores Reynolds's triangulation of contemporary understandings of reproductive biology, the socio-political limitations of primogeniture, and the supposed medicalised immorality of gender and sexuality non-conformity. Reynolds, I argue, uses his own frequently patriarchal and strangely conservative readings to undermine the very patriarchal and often politically conservative class system he ostensibly loathes.

A. Boucher, *Science, Medicine, and Aristocratic Lineage in Victorian Popular Fiction*, Palgrave Studies in Literature, Science and Medicine, https://doi.org/10.1007/978-3-031-41141-0_3

Though rarely read today, G.W.M. Reynolds was one of the early Victorian era's most popular authors whose writing 'was widely pirated, plagiarized and imitated'; his obituary in *The Bookseller* in 1879 even called him 'the most popular writer' of his time.[3] Though he wrote and published prolifically (with his potential use of ghost writers leading to some later attribution difficulties), his most widely read texts were serialised from 1844 to 1856: The *Mysteries of London* (which will, going forward, be abbreviated as *MoL*) and its prequel—the focus of this chapter—*The Mysteries of the Court of London* (abbreviated as *MoCL*); his particular brand of Urban Gothic penny fiction was equal parts silver fork novel, Newgate novel, radical Chartist propaganda, and soft-core pornography. Penny fiction (also called penny bloods or penny dreadfuls) is a broader category of regularly serialised fiction (often but not exclusively Gothic in nature) which was aimed at working-class readers and therefore cheaply produced and generically grouped by their penny price tag. For most of the nineteenth and twentieth centuries, penny fiction was

> all but forgotten, a mythical creature barely mentioned as something comparable to the monstrous hybrid of gothic novels and better forms of serialized popular fiction. This rather unkind perception stemmed from the original Victorian middle-class viewpoint on the penny bloods, which perceived this subversive literary form—violent, licentious, almost freely available to the working-class and, most of all, beyond their control—as dangerous,

although rediscoveries of established penny fiction and new discoveries of preserved copies of presumed-lost texts have enabled, over the last few decades, a critical reappraisal.[4] The scholarship of Anne Humphreys, Louis James, John Springhall, Helen R. Smith, Robert L. Mack, and Edward S. Turner, among many others, have opened up considerable insight into highly popular and prolific literature which, by virtue of its working-class audience, was not deemed worthy of consideration. Much Gothic and penny fiction was deeply concerned with classed medical and safety issues (as Anna Gasperini discusses at length in her work on the Victorian Penny Blood and the 1832 Anatomy Act); more intriguingly, it was considered a pathologising force in and of itself. This is of course true of most popular or genre fiction, which in almost all of its eras and forms has provoked medical concern or outrage over its supposed effects on a reader's delicate system (we will see this again as an explicitly stated medical concern in the next chapter on sensation fiction). Penny fiction was put in decidedly pathological terms by the sociologist and journalist James Greenwood in his 1874 *The Wilds of London* where he attacks penny fiction as 'a plague [...] bringing death and misery unspeakable', in the same league as 'the cholera and cattle plague', and with G.W.M. Reynolds himself as infectious and 'cunning as the fabled vampire'.[5] Though most penny fiction was viewed by middle- and upper-class critics as potentially dangerous to the delicate and influential psyches of the working classes—with the genre's reliance on sensational stories of gore, crime, seduction, and the supernatural—it is *MoCL* which Greenwood targets for his particular ire, likely because of the author's radical Chartist roots which crop up more vitriolically in this text than in his others and which make his most direct attack on aristocratic systems, through the vehicle of aristocratic fertility. Fertility as a metaphor for monstrosity, identity, and hybridity was nothing new: Diana Pérez Edelman discusses at length in her *Embryology and the Rise of the Gothic Novel* (2021) the extent to which Gothic fiction was concerned with fertility, conception, gestation, and procreation.[6] But it is the combination of Reynolds's 'infertility as moral and biopolitical bankruptcy', his Chartist politics, and (more likely than not) his bawdy content which provoked such ire.

MoCL revolves around twenty years in the life of George IV during his days as Prince of Wales and Prince Regent; although *MoL* is more famous in Victorian scholarship, *MoCL* focuses far more heavily on the lives of the aristocracy, as well as expressing more vitriol about leadership through aristocratic bodily dynamics than *MoL* does.[7] Reynolds, who had a background in Chartism and French Republicanism, uses this light pornography to slip a major part of his political agenda through the back door: to unmask aristocrats as unhealthy voluptuaries ill-suited to the government of a nation:

> By the living God, all this [aristocratic injustice] is intolerable [...] it assuredly is far more than sufficient to make ye chartists, republicans, and communists [...] But, no; the working men of England require not sophistry [...] to account for the evils which they endure. The causes are too palpable, too glaring, too apparent [...] for the causes thus alluded to exist in [...] chiefly our aristocracy, with its hereditary titles and its law of primogeniture, its usurpation of all the governmental and administrative powers of the state, its heartless tyranny and its cold-blooded avarice, its voluptuousness and luxury, maintained at the expense of starving millions.[8]

Chartism was a working-class movement that emerged in the late 1830s in Britain, with roots in and overlaps with other revolutionary or political movements on the Continent and in America. It resulted in the mass-protests of millions of British citizens seeking radical reforms and comprehensive restructurings of socio-economic and political systems. Chartist aims included universal male suffrage, secret voting ballots, the removal of property qualifications for, and the payment of, MPs (to ensure working-class men could be elected), equal constituencies, and an annual parliamentary election to allow the ousting of MPs proved to be corrupt or inefficient. In short, its goal was to remove the mechanisms by which aristocrats and other high-ranking citizens could embed themselves in the most important structures of influence. Most importantly for this chapter, Chartist thinkers also looked backwards to look forward: to the 1215 *Magna Carta* from which their new 1838 *carta* took some inspiration, to ancient or medieval concepts of leadership, to 'deep local and historical allegiances', and to the political and economic considerations that came with the sixteenth-century development of Anglicanism.[9] Reynolds used this Chartist idea of political lineage in a much more literalised way: in *MoCL*, Reynolds looks back on the morally corrupt aristocrats who existed a generation and a half before—and had presumably only just managed to produce—the silly, sickly aristocrats of silver fork fiction of the previous decades, who in turn produced and benefitted from the terrible working-class conditions of his age.

Michael Diamond argues that 'Reynolds's readers seem to have enjoyed both his revolutionary diatribes and the sex and violence; it is not clear which his detractors found more deplorable.'[10] But undoubtedly not all of Reynolds's readers agreed with his political views, nor sought out the texts for that purpose; in fact, some of Reynolds's most enthusiastic readers and collectors were part of the aristocracy themselves, like the first Baron Queensborough who proudly affixed his coat of arms to his bound volumes.[11] That said, Reynolds's many publications, including his *Reynolds's Newspaper* which 'stood alone as the most popular and stable radical weekly', was largely aimed at a working-class audience.[12] Reynolds's work even makes an appearance in the work of English journalist and reformer, Henry Mayhew, who reports in his *London Labour and the London Poor* (1851) an incident of working-class people reading Reynolds aloud and 'cursing the aristocracy'—a regular social event for many of the newly literate lower classes, who found enjoyment and novelty in literature written specifically with their views and interests in mind.[13]

The dynamics within Reynolds's texts are often at odds with his very staunch political perspective; this may account for his aristocratic fans despite his clear working-class target audience. His more notable contradictions, which form the basis of this chapter, include his willingness to revel in what he deems to be sexual looseness and gender non-conformity and to use those prurient depictions to castigate such promiscuity or profligacy as a class failing. And although Reynolds rails against any inherited status, his plots frequently offer titles and wealth to his moral, middle-class characters as the reward for good behaviour. This paradoxical reward system implies both that leadership and status should be earned individually, and that inheritable power is the ultimate prize: that those who morally *can* procreate *should* be given priority in a system reliant on procreation, despite that system eventually rendering its participants *unable* to procreate. Rohan McWilliam, historian of Victorian labour, succinctly summarises these innate contradictions by saying that, to Reynolds and his readers, the aristocracy was 'the one group in society that is perceived as truly free': while freedom on one hand implied glamour and empowerment, it also spoke of aristocratic independence from feudal responsibility and a denial of the obligations they owed to or greater influence their system had on the populace.[14] This simultaneous reinforcement and undoing of the cultural hegemony of the nobility, even in the most radical anti-aristocratic literature, becomes even more muddled when Reynolds engages in nineteenth-century discourse on fertility studies.

Medicine, Morality, and Masculinity

Literary critic Len Platt argues that the medicalisation of aristocratic male characters in Victorian literature (who are usually pathologically indicted through their sexual diseases, gout, and poor mental health) was merely a common trope or 'narratological hoop' through which the characters were jumped in order to demonstrate the moral failings of the upper classes.[15] While Reynolds certainly utilises all of these clichéd 'class' illnesses as what he considers to be evidence of the moral failings of the aristocracy, his usage and coding of the aristocratic body and lineage is far more complex than Platt allows. While Reynolds was not the first author or political activist to portray aristocratic bodies and heredity in a negative light, his contributions to this trope adhere to a significant pattern in the larger discourse of early nineteenth-century perceptions of aristocratic biopolitics, in which the 'regulation of biological processes and functions became increasingly important to policy makers and public health campaigners over the course of the nineteenth century'.[16] This is especially true at the time of Reynolds's writing, in which, as David Rosen argues in his work on Muscular Christianity, policy and power was directly tied to medicine, biology, and masculinity: the 1832 reform bill, the Chartist movement, European revolutions, and the British primogeniture debates of the 1830s and 1840s often led to 'the subject of gender and masculinity, for the debate over who should rule often devolved into a debate over who belonged to that privileged group called "men"'.[17] Aristocratic bodies, which had previously been literary conduits for discussions of performative pathology in the silver fork novels, here become a battleground on which morality, medicalised masculinity, and primogeniture can intersect.

Of all Reynolds's varied arguments against the aristocratic establishment, one of his subtlest and most complex points of attack is found in his focus on impotence and infertility. Children are conspicuous by their absence from the text. Until the very end of the series, none of the dozens of aristocratic characters can produce a single child in wedlock. Though many illegitimate children are begotten by both male and female nobles, they system of primogeniture only qualifies children by their legitimacy, with legitimate *male* children being the surest means of the line's survival and the most definite proof of the masculine virility of their father—which is a special preoccupation for Reynolds, who constantly reinforces that infertility or impotence lies with the male partner. In part, this portrayal mildly absolves female aristocrats who are exploited by the patriarchal nature of the aristocracy, although ultimately Reynolds's attack on aristocratic male fertility and supremacy occasionally reifies those very patriarchal Victorian mores around masculinity, effective leadership, and control that he seeks to undermine. Infertility inside of wedlock is presented by Reynolds as a badge of both immorality and ill health, very much linked in the manner of the 'king's two bodies'. Despite his contradictions and nakedly manipulative ideology, Reynolds underscores some very real and practical concerns that the general population would have about aristocratic infertility. Many of his readers would still be able to remember the various succession crises from 1817 to 1837 which were brought about by the fertility issues of George III's children and their spouses, the frequent changing of monarchs, and the political and economic instability this brought with it.[18] It is against this background that Reynolds presents his argument: that due to the voluptuous excesses and gender non-conformity of the aristocracy (voluptuous excesses which

Reynolds himself wallows in to boost sales), rule by primogeniture cannot work at its most basic, biological level and should be eradicated from the political system.

While Reynolds's depictions of male infertility are medically speculative and rooted in his own concepts of morality (rather than borne out of his own dogged scientific understanding), he is to some extent engaging with actual scientific work on aristocratic heredity and sterility which (as we will see in the next three chapters, percolated throughout the entirety of the nineteenth century and beyond). Erasmus Darwin, writing his *The Temple of Nature, Or, the Origin of Society: A Poem, with Philosophical Notes* in 1803, hypothesised that upper-class heirs were often so only because all other family members had died off or failed to reproduce, that they were 'not unfrequently the last of a diseased family'; he locates much of this culling 'disease' in indulgences of all kinds—specifically citing alcohol use, but including other vices which could kill off a family line, such as eating too much salt.[19]

Only a few years after *MoCL*'s serialised run finished, the famous Victorian surgeon and fertility researcher T.B. Curling reported (in a tone that matched much of Reynolds's moralising) that 'sterility oftener depended upon males than females' and that 'a man who is unable to fulfil the command, "to be fruitful and multiply" is [...] periling the happiness and perhaps health of a woman'; men so 'incompetent to their marital duties' are candidates for divorce.[20] As late as 1892, *The British Medical Journal* was still dancing around the same issues that Reynolds hypothesised over in the late 1840s, stating it was a well-known fact that 'aristocracies and families living in luxurious social conditions do not habitually keep up their numbers'.[21] It is important to note that this chapter is not about Reynolds actively engaging with real scientific and medical studies around aristocratic fertility, but rather using them as set dressing. Reynolds is far more concerned with now-discredited ideas like biopolitics or biological essentialism, as it concerns gender and sexuality, to make a political case against the aristocracy. That said, although Reynolds uses scientific issues as a servant *of* literature rather than being in dialogue *with* literature, his way of writing about them was also not (as we have seen) totally disconnected from medical and scientific rhetoric. Historian Stephen Halliday cites that moralising, 'while irrelevant to the problems at hand, was a characteristic tendency [of] many well-intentioned medical practitioners of the time'.[22] Although Reynolds is perhaps less concerned with the *actual* discussions being had by Victorian scientists and medical practitioners of his day, this chapter acts as a fulcrum on which genre fiction's more casual treatment of aristocratic bodies begins to transition into a far more informed, engaged, and even anxious representation. The silver fork novels traced which medicalised actions *positively* indicated a true aristocrat, and how those actions evolved in the next generation through *habitus*; this chapter traces what medicalised actions *negatively* indicate a true aristocrat, and how those will actions destroy that system in the next generations. The action-to-essentialism moral pipeline that Reynolds sets up here will go on to strongly inform the next three chapters which, with increasing dread, question the future viability of the aristocratic apparatus via its past actions, and what lessons the rest of humanity can learn from aristocratic pathology and lineage.

The bigger point of critical engagement in this chapter is how Reynolds portrays ideas of masculinity in relation to fertility. Though Reynolds's use of gender and sexuality in *MoCL* is complex, it is not necessarily sophisticated. His plots and extended bodily tropes indicate a confusion of biological function with contemporary cultural mores, and he provides no definitions nor adheres to any strict word-choice in his rhetoric on the body, gender, and sexuality. Since he relies on reproductive biology as a baseline for subjectively calculating health, normativity, and social suitability, quotations from *MoCL* may contain terminological overlap; however, in the framework of my criticism on Reynolds, I have applied a strict biology/culture schism between the terms 'male' or 'man', and terms such as 'manhood', 'unman', 'manliness', and 'masculinity'. The former implies a biological classification—at least in the way those terms were deployed in the nineteenth century—while the latter implies a set of cultural ideals. The argument becomes especially muddied in Reynolds when he toys with ideas of cross-dressing, non-binary identities, or other types of nonconformity, although almost all of these in *MoCL* are negatively portrayed as violating a gender and sexuality binary and thus rendering a character neuter (as will be discussed in depth below). Reynolds is more concerned with gender *ideals* as they relate to class, and does not operate with any kind of genuine engagement or good faith in terms of queer identities. There is ample need for further discussion of queerness in Reynolds but, for the purposes of scope, I will limit my discussion to mainstream nineteenth-century understandings of gender and sexual performance as it relates to Reynolds's ostensibly cis, straight characters.

While Reynolds's biological and queer terminology is not clearly defined, the class-based models of masculinity that he celebrates or disparages *are*. Reynolds venerates working-class men who '[r]ise early, [and] toil hard all day', while he abhors 'the pampered, insolent, overbearing aristocrat'.[23] James Eli Adams explores in his influential work on Victorian masculinity, *Dandies and Desert Saints* (1995), that the idealised roles of manhood in the Victorian era included but were not limited to the 'gentleman, dandy, priest, prophet, soldier, and professional'.[24] There is, of course, no single, unified Victorian concept of what it meant to culturally embody one's sex. Manhood could be in contrast to womanhood, boyhood, or animality; for Reynolds, manhood was in contrast to all three. Reynolds's texts are purveyors of Republicanism, lower- and middle-class morality, and the Protestant work ethic. As such, they subscribe to a type of heteronormative masculine identity which was best summarised by John Ruskin—though there is otherwise no connection between Ruskin and Reynolds—in 1865 essay, 'Of Queen's Gardens': that the 'man's power is active, progressive, defensive. He is eminently the doer, the creator, the discoverer, the defender. His intellect is for speculation and invention; his energy for adventure, for war, and for conquest.'[25] This is a vision of masculinity which many critics consider characteristic of the Victorian period: that the 'Victorian period registered the most extreme form of gender segregation yet seen in an industrialized nation'; that 'self control, restraint and distance became the hallmarks of ideal masculine identity'; that 'the meaning of masculinity was self-evident and it involved emotional reserve and physical discipline'; that the 'healthy man is strong, assertive, tolerant, moderate in his appetites, hard-working, adventurous, responsible, and wise'.[26] The aristocracy (frequently portrayed in literature and art as leisured, concerned with fashion, and of immoderate appetites) was sometimes seen as lacking in masculinity, a trope or perception that Reynolds exploited in his own literature for his personal politics. In particular for Reynolds's working-class audience, the aristocracy in general, and certainly in Reynolds's representations of them, would have been seen as effeminate.[27] Despite his radical history and championship of the working man, his ideals of masculinity are fairly conservative, even for the time, and all largely based in notions of middle-class manhood.[28] This is, of course, only one-half of a binary set up in Reynolds's discourse on gender. He does address biological and medical issues surrounding his female characters as well. However, since his treatment of femininity is equally complex, but less concerned with fertility, transgressions of the body, and Republican politics, I will discuss the other half of this binary only as it directly concerns male reproduction.

Reynolds challenges the manhood of his aristocratic characters and deprives them of fertility through a combination of two processes: 'feminisation' and 'emasculation'. Reynolds does not name them as such, but the processes are very distinct in his work. Feminisation occurs when Reynolds applies physical and behavioural traits that he considers feminine to a male character: physiological frailty, what he defines as female beauty, vanity, and a willing passivity. His process of emasculation is characterised by the symbolic neutering of a male character by depriving him of that which Reynolds considers masculine: sexual vibrancy, agency, strength, and hardiness. Feminisation adds traits, emasculation subtracts; the former creates a character with the capacity to be either gender, the latter neither. Both are utilised for the same end: to imply a disrupted or disturbed physiology which has a direct impact on one's health, reproductive potency, and suitability to rule. Though Reynolds's aristocratic male characters are often confused and wavering about the state of their bodies and masculinity (and therefore its connection to politics, health, and morality), Reynolds is confident in his own cultural constructs, foiling his own male characters by being so tonally assertive. Gender theorist Harry Brod, working from sociologist Michael Kimmel, sheds light upon Reynolds's treatment of gender norms as clearly defined, saying that 'for a man to admit that he has questions about masculinity is already to admit that he has failed at masculinity'; Reynolds seemingly has no questions, though his characters do.[29] While John Tosh argues that definitions of bourgeois masculinity in the nineteenth century were in no way certain or unified, in Reynolds's work there is no room for ambiguity, uncertainty, or vacillation when it comes to gender and, by Reynolds's extension, health, lineage, and political power.[30]

Though Reynolds plays with notions of gender, sexuality, and identity through the bulk of his works, there are three characters in *MoCL* which represent three distinct junctures within his arguments about aristocratic fertility and gender polarity: Lord Florimel, the Earl of Desborough, and the Prince Regent (the future George IV). The lives of these men and their inability to produce children in wedlock characterise the potential outcomes of the aristocratic life-cycle, as perceived by Reynolds. He uses their fates as proof of the validity of his Republican politics, which argues strongly against inherited power, since '[d]epravity would seem to run in their blood, and to be as traditionary as their titles and estates'.[31] By using an overlap between gender performance and fertility to critique the aristocracy, Reynolds is able to manipulate his reader's assumed conventional mores regarding family values, gender binaries, and bodily norms into a more nuanced political argument. Jennifer Terry and Jacqueline Urla argue in their work on body deviance that there was a strong nineteenth-century belief that one's moral character was rooted in biology, which led to society's 'feverish desire to classify forms of deviance, to locate them in biology, and thus to police them in the larger social body'.[32] The following sections on feminisation and emasculation, therefore, explore forms of deviance in the aristocratic male body, how deviance affects fertility, and how Reynolds classifies and polices these individuals in the larger social and political body.

Feminisation

Lord Florimel represents the feminised male aristocrat in *MoCL*. He is one of Reynolds's least gender-polemic males, being a handsome dandy who takes his sexual pleasure by dressing as a woman, 'Gabrielle', in order to gain the trust of and then seduce honourable women in what is ultimately a twofold queer exchange: '"[W]e will be friends, bosom friends, Gabrielle, will we not?" "Till death!" Replied the nobleman. "And now let us seal our friendship with a kiss."'[33] To Reynolds, this predilection makes Florimel not only an exploitative cad, but also a sexual deviant. Apart from denoting homosexual tendencies (which may overlap, but are unaffiliated, with transvestism or transness—a distinction which Reynolds does not and could not make), Florimel's cross-dressing also conjures thoughts of lesbianism, since he is performing femininity during the sexual conquest of another woman.[34] This performance makes the heterosexual character register to a reader as a homosexual twice over, and is therefore twice as unable to produce a child (despite this actually being a cis, heterosexual encounter).

His duality exists right on the surface, beginning with his names: 'Florimel' is his ancestral surname which should be given to the sons who will continue his line, but 'Florimel' also means 'honey-flower' and is the name of a female character in Spenser's *Faerie Queene*. His first name, Gabriel, should indicate that he, like the archangel, is a harbinger of the births of important men. However, if one takes the biblical reference to its logical conclusion, the archangel is incapable of producing offspring of his own and merely announces the arrival of a moral, lower-class man who will rise to a position of leadership. Florimel feminises his male name by the addition of feminine qualities onto the masculine base: Reynolds specifically draws attention to Florimel's addition of extra letters to the pronunciation of his name: 'with that stress upon the final syllable of the Christian name [...] "Gabrielle Florimel," said the nobleman [...] laying a stronger emphasis on the "el."'[35]

Reynolds does not hesitate to locate this deviance far beyond Florimel's behaviour and in Florimel's body itself. The depictions of Florimel's physicality are almost caricatures of feminine beauty:

> [A] razor had never touched his cheeks, which has all the damask and peach-like loveliness peculiar to the softer sex. His complexion was singularly fair, clear, and stainless; his nose was small and perfectly straight, his lips were red and full, and his teeth brilliantly white and faultlessly even. His neck was long and gracefully turned, his ears remarkably small and delicate. He wore his rich chestnut hair flowing in a wavy mass over his shoulders; and as it was parted with great precision above the high and open forehead, its arrangement completed the feminine appearance of the youthful noble's countenance [...] For beautiful he indeed was,—not handsome [...] Florimel was very short for a man [...] and nothing could exceed the delicate whiteness of his hands and the diminutive modeling of his feet. His voice corresponded with this feminine style of beauty.[36]

The long accounts of Florimel's delicate beauty are purposefully gender-ambiguous, which only emphasise their transgressiveness; in his first series, *MoL*, one of Reynolds's main plots involved an attractive young man who turned out to be secretly a woman: 'He was a youth, apparently not more than sixteen years of age [...] his countenance, which was as fair and delicate as that of a young girl [... was framed by his] long, luxuriant hair, of a beautiful light chestnut colour.'[37] Since Reynolds constructs many of the same ambiguities around Florimel in *MoCL* (even their hair is the same colour and worn in the same way), it is not immediately clear that Florimel is actually male; a reader would be justified in suspecting Florimel was another of Reynolds's women in drag. Given the lightly pornographic nature of the work, Florimel's gender-ambiguous deviance may even extended to and 'corrupt' the reader, who may feel prompted by Reynolds to feel sexual attraction towards Florimel; in this way, Reynolds's working-class readership can feel the direct effects of the transgressive aristocratic bodily upon themselves and its 'contamination' of their own morality.

By the time Florimel is introduced in the text, other beautiful women have already been described in identical language; of the Clarendon sisters alone, one of whom becomes Florimel's wife, Reynolds says, 'their complexions were equally fair and beautiful [...] Their foreheads were high and open, their mouths small, and with lips red and ripe as cherries, and their teeth of pearly whiteness.'[38] Describing Florimel as 'peachlike' is meant to further pervert his gender and fertility—by relating his good looks in terms of fruit, Reynolds subverts a common metaphor for a sexual and fecund woman, just as he did when describing the Clarendon sisters' lips as 'cherries'. Understandably, fruit imagery is also a common way to express fecundity. Florimel is described very traditionally as a fertile woman, though a reader knows he cannot bear children—but the text simultaneously admits that he will never father children, either. Early in the series Reynolds indicates that, since Florimel's roguery and depraved habits render him incompatible with a gender-normative marriage, his line would almost assuredly end with him. 'Possessed of immense wealth, and with no parents nor elderly relatives to advise him, he devoted all his time and all his thoughts to the pleasures of love', showing that Florimel too well enjoys his autonomy away from the pressures and responsibilities of the family unit.[39] He also lacks the loyalty, reliability, or maturity which *should* be requisite for marriage and the successful rearing of children; instead he is '[f]ickle, inconstant, and easily excited by a new and pretty face'.[40] Everything about Florimel revolves around transitory pleasure and transgression, in direct opposition to the wholesome and long-lasting happiness that Reynolds implies is found in gender-binary family life. Creating a clear cause and effect, Reynolds introduces Florimel as an extremely feminised character and then states, 'He was unmarried and likely to remain so; for the idea of linking himself to one woman was, in his estimation, something too dreadful to contemplate', the implication being that if Florimel married, it might impede his association with the other women in his life—both the ones he conquered and the one he performed.[41]

The redemption of Florimel's fertility becomes one of the major subplots during the first five volumes of *MoCL*. Having met the beautiful but stubbornly virtuous Pauline Clarendon, one of the protagonists of the series, he chooses to reject his empty aristocratic life and prove his middle-class masculine worth to her through a total realignment of his body and behaviour. When circumstances force him to don female attire again for the sake of an intrigue, he sustains a concussion and becomes an invalid for three days. The immediate and severe repercussions serve as a warning: sexual 'sickness' breeds physical sickness. That he calls himself 'Miss Plantagenet' during his final instance of transvestism further ties this behaviour to a self-destructive aristocracy.[42] From that moment, his character rebuffs all that is feminine—Reynolds's focus transfers from Florimel's looks to his actions, from a body coded as a feminine visual object to a body coded as a masculine source of action. Reynolds celebrates Pauline as one half of a gender binary, for her idealistic womanhood inspires Florimel into idealistic manhood: '"If I be thus changed, Pauline [...] it is your bright example that has worked so salutary an effect."'[43] Their relationship undergoes several tests, but he never falters in his new devotion to middle-class morality and they ultimately marry at the end of Volume 5.

Though he is rescued from his decadent lifestyle by their marriage, the damage to his fertility seems to already have been done. When Volume 6 begins, set nearly twenty years later, they 'remained unblessed by offspring', though the logistics of the plot would not have been impeded by the presence of children.[44] Reynolds is quick to blame his characters for their infertility, and the background evidence he provides makes Florimel culpable, instead of his wife, Pauline. Where Florimel is in fact the last of his line (indicating a hereditary struggle with fertility, as warned about by Erasmus Darwin), Pauline's sister gives birth to a daughter. In later volumes, Florimel makes this niece his heir, for lack of a better candidate: as 'Lord Florimel had no children of his own, he soon learned to love [her] as dearly as if she were his daughter'.[45] Her heiress status, along with the details of her aristocratic birth and upcoming aristocratic marriage, means that the niece is the last hope for the continuation of at least four noble lines. Her untimely death in the last volume means the complete extinction of those lines, of which Florimel's is chief.

Denying Florimel the capacity for reproduction, Reynolds casts a pall on the character's health. The childless life Florimel had predicted for himself before marriage became the life he could not alter after marriage. Since he and Pauline look similar, Reynolds implies a certain fruitless, masturbatory solipsism in Florimel's attraction to his wife. Further and more significantly, transvestism was punishable under the law, as it was associated almost solely with homosexuality, and especially with homosexual sex work.[46] Therefore, Jennifer Terry's argument in her work on nineteenth-century science and homosexuality can easily be applied to Florimel, whose masturbatory relationship and early transvestite (read: homosexual) activities were both seen as acts of 'self-pollution [which] drained the male body of its vitality and left no offspring to show for it [It led] to a point of no return, leaving the 'youthful sinner' [...] in a state of 'physical impotence' that made an adjustment to heterosexual relations impossible'[47] To

Reynolds, deviant behaviour is inextricably linked with one's biological make-up, and immorality is a form of incurable sickness; one can be socially redeemed, but physiology cannot be amended.

Emasculation

On the opposite end of Reynolds's gender-and-bodily deviance spectrum is the Earl of Desborough, the most clearly emasculated character in the series. He is also the only character around whom Reynolds centres a frank and largely non-symbolic discussion about reproductive issues. Most other characters' infertility is only alluded to; for example, two noblemen who have proved incapable of begetting children in wedlock are described as frequent smokers, constantly 'impregnating the air with the smoke of their cigars'.[48] This ephemeral, impermanent impregnation is the only type they can generate with their phallic cigars which, by their very nature, diminish into ash. However, the Earl's situation is described in language of remarkable clarity as 'the lamentable physical misfortune which rendered me unfit for marriage, well knowing, in fact, that ten thousand sources of misery would eventually be summed up in the terrible word impotency'.[49]

Unlike some aristocrats in the series (including the Prince Regent, Letitia Lade, the Duchess of Devonshire, and the Marquis of Holmsford), the Earl of Desborough likely had no single real-life counterpart (there is no earldom associated with the town of Desborough), though Reynolds may have had an historical basis for the character. He may have been referencing the Earldom of Desmond, which went extinct three times in rapid succession in the seventeenth century, before ultimately being swept under the umbrella of a grander title and then largely ignored.[50] While the language used to depict the Earl's situation is non-symbolic, his character and body nothing but symbolic, making him the standard bearer for all emasculated male aristocrats in Reynolds's fiction.

Much like Lord Florimel, the Earl's entire existence is viewed as a vehicle for producing offspring: when production fails, his political existence fails. Unlike Florimel, however, the Earl is not infertile because of feminisation, but because he is medically impotent. The Earl's inability to participate in sexual activity denies him status as a man and keeps him embedded in what Reynolds portrays as a perpetual boyhood. The cause of the impotence is never revealed, though it appears to be a congenital issue or symptom of a childhood disease: he does not recall a time without it, and when he asks the physician, '"Then there is no hope [...]?"' the answer is, '"None, my lord."'[51] He is the only character seen to consult a physician about a non-life-threatening issue, and the only character whose physician is completely unable to provide any treatment. Since all other appearances of doctors in the series involve either childbirth or impending fatality, Reynolds singles out the Earl and traps him somewhere between life and death—repeatedly called a 'corpse', and yet still living.[52]

Much of Reynolds's frankness surrounding the Earl is composed through the Earl's own cognisance of his medical issues, a self-awareness which Florimel does not share. This knowledge, and the Earl's inability to move either fully into bodily life or bodily death prompts an anxiety-ridden break-down while Florimel and other aristocrats remain happily ignorant of the medical implications Reynolds writes into their lives. Where other aristocrats are vice-ridden automatons, driven solely by personal desire from one scandal to another, the Earl's quiet self-perception serves as a figurative first awakening of the titled class. As the Earl contemplates his purpose in society and realises he is incapable of fulfilling that purpose, his only recourse is to immediate self-destruction.

Michael H. Shirley states in his work on Reynolds, 'The solution to centuries of stagnation was not, he [Reynolds] believed, violent revolution to create a classless society, but a *peaceful* and yet constant agitation to bring about fundamental change.'[53] The Earl's self-destruction, while a violent act in itself, can be read in the context of the serial as being representative of the gentle transformation Reynolds hoped to enact in society, in total opposition to the revolutions occurring across Europe at the time of his writing. By awakening the aristocracy to their own truth and exposing them as no longer capable of fulfilling their socio-biological purpose, they might readily dissolve their own establishment. Though the Earl's suicide seems like an extreme action which is anything *but* peaceful, Reynolds uses it to demonstrate a willingness on the part of the aristocrat to do what he, Reynolds, believes to be right and to end their 'lives' as aristocrats on their own terms, instead of by bloody overthrow. Though infertile the nobles may be in his work, Reynolds's use of rehabilitated or virtuous aristocrats like the Earl and Florimel shows a surprising level of faith and optimism in the class he belittles, especially given how irrevocably he portrays their inbred immorality to be.

The key difference between the Earl and Florimel is that the Earl's condition is not based upon behaviour affecting the body, but is rather the result of the body affecting behaviour. He and, by extension, his position and estate are infirm and not self-sustaining. There is no insinuation that the Earl's condition was the result of decadence, for he is presented as a decent, if pathetic, character. His lamentation that he 'was madman enough to think and to hope that there might be such a sentiment as a love of divine nature [i.e. romantic love], apart from gross enjoyment, and existing rather as an essence than a sensuousness' is in direct contrast with the profligate speeches of other aristocratic males, but it serves the same function.[54] The Earl reveals a characteristic that renders him unsuitable for fathering offspring. Where Florimel once rejected standard family life for the duality of feminisation, the Earl wants embrace family ideals but does not have the capacity of even *one* sex to make it a reality.

The inverse relationship between the Earl and Florimel continues: Florimel's dissoluteness causes infertility, while the Earl's infertility causes dissoluteness. This connection is established early in the series when the reader is introduced to his wife, the Countess. She, being 'at times devoured by desires and rendered restless by fierce passions', is furious that he has not been able to consummate their relationship after years of marriage and refuses to participate socially as a wife inside the family unit: 'a cold, imperceptible tremor swept over her frame the instant that the earl appeared upon the threshold of the apartment'.[55] Wracked with guilt for his inability to satisfy her, the Earl encourages and even arranges for his wife to take a lover. Though he is sickened and further emasculated by the idea ('there was a profound melancholy devouring the nobleman's very vitals [...] his cheeks were growing pale, his form emaciated'), his wife's happiness and the need for an heir, legitimate or otherwise, are too strong an inducement to resist, further breaking down the idea of the aristocratic family into the cynical combination of mere alliance and appearance.[56]

In his book on body politics, Dominic James posits that in Victorian England, 'the rational mind was gendered male and the dependent body as female'.[57] This is a model which encapsulates the dysfunctional marriage between the Earl and Countess. She, 'whose passions were, however, more potent than her reason' is a slave to her bodily appetites and relies upon the Earl's strength of character and reason to keep her from infidelity.[58] Since his emasculation renders him passive, he has no strength of character, and the rational masculine mind is overpowered by the irrational feminine body: 'And if hell's flames were immediately to follow the consummation of her [the Countess's] frailty, she would not resign these few moments of Elysium to save herself from that eternity of pain.'[59] He fails to fulfil his part of the gender binary, leading to imbalance, domestic havoc, and moral erosion.

As with Florimel and Pauline, Reynolds again gives hope of a happy ending before ultimately destroying the lives of his sympathetic aristocrats. The countess repents, reconciles with her husband and together they undergo a moral, physical, and marital rehabilitation. However, the Earl is never able to recover from his shame and from the dishonour he allowed his wife to pursue. The constant fear of his impotence being made public drives him to despair and, combined with his reflections on the futility of his existence, he commits suicide. Being virtuous, the Earl is granted the only significant moment of aristocratic agency and masculinity in the series; since the Earl is noble, in both the literal and metaphorical senses of the word, he uses this moment for the betterment of the people by resigning his 'faulty' aristocratic body. Throwing himself from the roof of his stately home in the presence of his wife, the manner of his death parallels his plummet in her eyes, as well as the unstoppable descent of his family and class. He prefers a swift death to a long life filled with the knowledge of his shortcomings. His last words, 'You will make the world believe it was an accident, Eleanor', implore his wife to maintain their public face and do the best she can for their class, even has he consciously acts against the living façade propagated by the aristocracy.[60] To the last, he cannot bear the indignity of a revelation and must maintain the appearances required of his station.

Reynolds refuses to give most of his aristocratic characters the joyous and fulfilling resolution he begins to set up, depicting the aristocracy's collapse as inevitable. That he creates tragedy more frequently for his *sympathetic* aristocrats—men he described as 'charitable in the extreme' and 'affable and gracious', but 'ill-fated'—underlines the harshness of the aristocratic institution, which makes victims of both its members and the lower classes it oppresses.[61] While Reynolds certainly expresses cathartic pleasure in the fictional downfall of an unpleasant leader, the destruction of his sympathetic aristocrats is the true means by which he advocates change.

Feminisation and Emasculation in Unison

As the core antagonist to a revolving series of protagonists, the Prince Regent suffers the brunt of Reynolds's criticism: not only does his status as future king attach the most serious political ramifications to his infertility, but he also embodies emasculation and feminisation in equal parts. The Prince was feminised in popular culture—the playwright Richard Brinsley Sheridan said the Prince had '"the most womanish mind" he had ever come across' and the Duchess of Devonshire reported that he was 'too much like a woman in men's cloaths [*sic*]'.[62] This feminisation was partly due to the Prince's adherence to the model of the dandy. The dandy was, by the time of Reynolds's writing, falling deeply out of fashion along with the silver fork novels and becoming what Adams calls a 'grotesque icon of an outworn aristocratic order, a figure of self-absorbed, parasitic existence'.[63] Danahay goes so far as to say that 'being a dandy was about as close as any man could come to rejecting his masculinity'.[64] The Prince Regent's admission in *MoCL* that 'I was formed and fashioned to spend my existence pleasurably, and not in the routine of business and serious affairs' seems to be roughly, if not totally, founded in truth: the Prince Regent was reputed to blatantly prioritise pleasure over his obligations, in clear contrast to the masculine Protestant work ethic which Reynolds espoused for his readers.[65]

While the term 'Protestant work ethic' is anachronistic to Reynolds and his work, being first coined by Max Weber in 1905 and used by him to understand the economic differences between countries that are predominantly Catholic or predominantly Protestant, the concept greatly underscored Reynolds's rhetoric. Weber writes that '*one's duty consists in pursuing one's calling* [*Berufspflicht*], and that the individual should have a commitment to his 'professional' [*beruflichen*] activity', a notion that Reynolds applies to his characters by rewarding those who carry out their professional duties and punishing those who do not.[66] Conceptions of the Protestant work ethic for Victorian scholars are often coupled with Samuel Smiles's 1859 *Self-Help* and have become synonymous with middle-class labour, asceticism, and respectability, especially in opposition to perceived upper- and lower-class sloth, immoderation, and vice. Danahay goes so far as to argue that '[m]ale Victorian identity was modeled on the Protestant work ethic', while Adams qualifies this argument more, stating that self-discipline 'is the distinguishing feature of professional men' and that Victorian gender tropes were informed by 'the religious paradigm of Victorian self-regulation'.[67] It is this combination of feminine dandyism and the lack of masculine work ethic that enable Reynolds to target the Prince with ease.

In *MoCL* the Prince Regent is first introduced as a feminised character. He is in delicate health (being severely hung-over) while gingerly attempting a long bath and toilet at his vanity table. The implication by Reynolds is clear: immoderation breeds weakness, and weakness is womanly.[68] The Prince is interrupted by Lady Letitia Lade, 'the Amazon'. Though a real-life friend of the Prince Regent, her portrayal in *MoCL* is far from biographical; rather, she is a mirror image of the Prince, her masculinity underscoring his femininity. Letitia revels in her marriage to a lord who 'is well-nigh in his dotage [...] He lets me do just as I like', and she repeatedly takes advantage of her husband's frailty, as well as the weaknesses of the aristocratic men in her circle.[69] She dresses in men's apparel in contrast to the Prince, who is still in his dressing gown. The Prince's dress immediately places him in the confines of the feminine body as defined by Reynolds: part of the pornographic element in the text revolves around beautiful female characters being voyeuristically presented to the reader in an early-morning state of undress, and these scenes were emphasised as essential moments in the text by the illustrations which accompanied each volume.[70] As the Prince sits in bed, indecent but for the bed sheets, Letitia says, "'[H]ave your bath, by all means. Here, I will give you your dressing-gown and slippers" [...] "And you mean me to rise in your presence?" asked the prince.'[71] When he does, she lasciviously leers at him for being '"in *dishabille* [*sic*]," she added, with a significant glance at the prince's figured silk dressing-gown and embroidered red morocco slippers'.[72] The Prince's introduction is also the first instance of the repeated dressing-gown nudity trope of the text, and he is the only man to join Reynolds's coterie of semi-nude women.

The Prince and Letitia soon draw back into the Prince's bathroom to consummate their relationship. He is vulnerable and frail, she is strong and well; he is undressed in the manner of other female characters, she is dressed as a man; for the sexual act, he retreats further into his suite while she moves forward, invading his space. He even compares his bathroom to 'the harem of a Turkish palace', a safe, appropriated living space solely for the female (in this instance, the Prince), and of which the male (Letitia) is only a visitor.[73]

This scene is crucial in the medical analysis of the Prince in subsequent volumes. By placing him in the confines of a weakened female body, Reynolds is able to construct a correlation between the Prince's fertility and venereal disease, specifically syphilis, which the text implies was perhaps inherited from the Prince's own father. The Prince, who does not have the desired masculine hardihood present in Reynolds's more admirable male characters, is trapped in an ouroboros: femininity leads to sickness, which leads to further femininity, which leads to further sickness. As the narrative continues and one sees the results of his many liaisons, the evidence of syphilis begins to accumulate, most notably that many of his sexual conquests have fertility issues after exposure to him. Mrs Fitzherbert, Letitia Lade, and Venetia Trelawney are never able to give birth at all, while Queen Caroline, Octavia Clarendon, and Agatha Owens each give birth to a single girl (all of whom die) and never conceive again.

More than fertility issues, his mistresses and children struggle with mental and physical health in a way that suggests syphilitic contagion. Agatha Owens gives birth to his stillborn child before dying in an asylum. Venetia Trelawney becomes 'a prey to melancholy'.[74] After his ruination of Octavia Clarendon, she goes mad, feeling his '"coils environ me!" [...] a terrible laugh which pealed from her lips spoke out the appalling truth. Octavia Clarendon was a maniac!'[75] She never fully recovers and dies young. Twenty years later, their illegitimate daughter encounters the Prince for the first time and grows madder with each new exposure to him, finally running from him in a frenzy and throwing herself to her death. '"Perdition!" ejaculated the prince. "She is mad! She will do herself a mischief!" [...] At this instant a terrific cry burst forth [...] Down she had fallen, down, down.'[76] His legitimate daughter, Princess Charlotte, is presented with an unspecified mental disorder which frequently gives her pensive bouts of melancholy and anxiety over her heredity, believing she came from 'a race whose infamies had rendered it accursed in the sight of Heaven, and whose punishment had to some extent, in the person of the lunatic king, commenced upon earth'.[77]

Even women who spurn his advances suffer from brief mental instability, as though they ran the risk of sexual infection through sheer proximity to him: the Countess of Desborough says he has a 'polluted embrace', Rose Foster 'felt as if she were going mad', and Pauline Clarendon's 'whole form shook as if with a strong spasm passing through it'.[78] The Prince's femininity is tied tightly to the concept of ill health, and the exposure of others to his feminine sickness leads to the contagion and destruction of those closest to him: his mistresses and children. By placing the Prince Regent in a feminine form, Reynolds takes arguments against the aristocracy into areas where typical political discourse could not tread as easily—namely, into an attack with a biological imperative.

The Prince Regent's femininity is in many ways the cause of his emasculation, since it traps him in the liminal space between the binaries of manhood and womanhood, making him perform as neither quite one nor the other. Reynolds treats his gender-atypical male characters as almost mule-like in their hybridity, unable to reproduce because of a perceived impurity or duality. While the Prince is certainly capable of the sexual act and precipitates several pregnancies, he is also presented as the anti-father, the destroyer of families and the next generation. Reynolds's preoccupation with fatherhood as a necessary component of leadership is seen most clearly in the Prince Regent, who is depicted as being capable of neither. This metaphor is at the forefront of Reynolds's Republican politics, portraying the Prince as 'the heartless man who is one day to become the Father of his People!', as well as literally, as a man who fathers sickly, still-born, or murdered children.[79] During a nightmare about all his sexual crimes against women, the Prince sees the

> lovely girls whom he had wooed and either seduced or ravished in his time, fair creatures who had gone down to the tomb with broken hearts and blighted affections. [S]ome of them appeared to have babes in their arms,—spectral babes, as ghastly as the parents […] babes which were the fruit of those pleasures that the prince had purchased either by means of the most insidious perfidy or the most heartless violence. And those infants had all died either at their births or soon afterward, some sacrificed to the fatal compression adopted by their miserable mothers to conceal their shame, others murdered outright by suffocation, or even by a bloody violence, during the puerperal aberrations of those dishonoured beings who gave them birth […] Yes, mothers and babes alike glared thus on the prince, babes and mothers reproached him equally with their dead, lustreless orbs [H]e was the man who deserved to be stigmatized as the moral murderer, if not the actual assassin.[80]

These deaths are an attack on Hanoverian rule: that the Prince, instead of providing fatherly nurture to his subjects, maintains his comfort, power and pleasure through the destruction of their innocent lives. He unmans himself through his refusal to accept the consequences and responsibilities that are the result of his licentiousness, instead looking for 'pleasures which are purchased by tears, lamentations, and premature deaths'.[81] The Prince is medically and socially neutered from producing legitimate children by his own physiological defects, decadent lifestyle, inability to provide for himself, and inability to provide for others. In Reynolds's work, the Prince and his line are untenable in the changing landscape of the nineteenth century.

Further emasculating him, Reynolds depicts the Prince as situationally impotent in several instances. One occurrence was based on the reports of the Prince's real-life wedding night with Princess Caroline of Brunswick.[82] Reynolds signals the importance of recalling such an event by pulling the narration away from the wedding party and asking the reader to 'resume the thread of our narrative in its proper place [...] the Prince of Wales was bearing home his bride to Carlton House'.[83] That the public's 'proper place' is with the newlyweds in their bridal chamber illustrates not only the stakes the nation had in their marital relations, but also the importance Reynolds places on aristocratic sexuality in the confines of his argument and the importance their bodies have as narrative devices. Reynolds reports that, despite the huge political importance of the conjugal meeting, the Prince's decadence overcame his responsibilities: he fell down senseless with drink and, come morning, 'only one person had lain in that nuptial bed'.[84] That he could not keep himself upright on his wedding night is a clear double-entendre, providing the punch line to Reynolds's long discourse on fertility and debauchery.

The Prince's virility becomes the butt of a second grim joke, this time centring on the Prince's prowess in the face of middle- and lower-class virtue. In what turns out to be an equally farcical and horrifying series of events, he begins kidnapping women who are unresponsive to his wooing and imprisons them in a secluded domicile with the intention of obtaining their favours through violence. He kidnaps women more than half a dozen times, and yet he never once successfully commits an assault; there is always an interruption or escape, as though the universe conspired to keep him from consummation: 'And that she would become his prey beyond all possibility of salvation or rescue, he did not doubt [and yet he became] thoroughly baffled by Camilla's heroic flight'; and again, in Vol. 3: 'as every moment saw her struggles becoming weaker and her cries more subdued, while the triumph of the prince appeared more and more certain. But suddenly the door was burst violently open, and Tim Meagles [the Prince's friend] rushed into the chamber.'[85] Though the daring escapes work as merely wishful triumphs of the lower classes over the abuses of the upper class, they also play into Reynolds's construction of the Prince as a sexual weakling: firstly by making him resort to such low acts, and secondly by making him unable to perform the acts, even when he is theoretically in total control and domination.

Reynolds's denial of aristocratic sexual dominance is reinforced by the Prince's close friend, the Marquis of Leveson, a childless man who is a 'widower, and already on the bleak side of sixty', who attempts violence against women in a similar manner.[86] No doubt inspired by the Prince's operation in earlier volumes, the Marquis converts a secret chamber in his mansion into a den of booby-traps to ensnare his victims, the most notable item being a chair with spring-loaded manacles 'that clasped her wrists and the steel bands that fastened their gripe [*sic*] upon her shoulders', rendering its user helpless.[87] Once again, however, every attempt is foiled by a last-minute interruption or complication, rendering the Marquis impotent: 'But just at the moment when the Marquis of Leveson fancied that our heroine was sinking into a profound insensibility [the door] of the suite was thrown violently open.'[88] That his lair is inside his home instead of in a separate location further corrupts the idea of a healthy, sexually normative home-life; it would be impossible for the Marquis to marry or rear children in a location which comprises such horrors, the two options clearly depicted as mutually exclusive. Being long widowed and childless, it is significant that the Marquis chooses to remain so and instead participates in sexual encounters which could only produce illegitimate children; it is, in Reynolds's logic, a complete rejection of an aristocrat's biological duty, and a failure of the aristocratic body.

The Marquis proves to be just as feminised and emasculated as the Prince Regent, who, at the head of the aristocracy, set the standard for the actions of its members. Just like the Prince Regent, the Marquis is feminised by his weakened female body, with its history of 'long and serious illness' and his dandified habits, 'with an admirable wig, a brilliant set of false teeth, dyed whiskers, the use of all the choicest cosmetics, and the artistic skill of a Parisian tailor'.[89] He is then equally emasculated through his bestial physiology, which stands in direct opposition to Reynolds's depiction of the thoughtful, genteel ideal of Victorian manhood. His inhumane actions against women, as well as his physical description, remove masculine characteristics and render him an animal: 'At the first glance he might have been mistaken for a bear escaped from the zoological gardens [...] for he wore such an enormous quantity of hair about his face as almost to destroy the features that identify him as a human being.'[90]

The Marquis's social position requires him to look to the Prince as an authority on all matters and to mimic the Prince whenever possible; further, the Marquis is part of the Prince's close social circle, the rest of whom knowingly accept their friends' repeated rape attempts. When the Marquis's attack on a young girl seemingly kills her, his friend debates reporting the Marquis to the authorities before shrugging it off, saying, 'I myself have not been immaculate enough in my life to feel justified in becoming the accuser of others [...] I have so many faults of my own to screen that I consider it but just to throw a veil if possible over the faults of my friends.'[91]

The Marquis serves as a stand-in for the Prince Regent. They are near the same age, have the same habits, friends, social status, physicality, and are both feminised and emasculated. While the history of the Prince Regent is too well-documented for Reynolds to punish him accordingly, the Marquis of Leveson did not exist and could therefore receive poetic justice for his decadent body and lack of gender polarity, without Reynolds rewriting history.[92] The Marquis's death, therefore, would be equally fitting for the Prince Regent: the Marquis is captured in his own booby-trapped chair while his mansion burns to the ground, in a literal and metaphorical inferno. As a crowd gathers outside to watch, 'a large portion of the building gave way, and much of the interior was for a few brief instants exposed to the view of the crowd gathered in the street', revealing the Marquis's horrifying chamber, true nature, and status as the final victim of his own decadent abuses.[93] The fate of the Marquis is Reynolds's compensation for his arch-villain, the Prince, escaping the serial unscathed. However, Reynolds implies that his serial was to the real-life Prince what the fire was to the Marquis: a force which strips away the glamorous surface to reveal the moralistic truth to the general populace. Reynolds writes, 'And then a man will arise [Reynolds's allusion to himself], bold enough to tear away the glossy veil which hides the deformities of the mighty by birth.'[94]

Conclusion

Trefor Thomas argues that Reynolds's 'weekly penny fiction can be understood as an impure, almost hybrid mode, half weekly newspaper, half romance'.[95] While his weekly fiction did include elements of contemporary news stories, *MoCL*'s outlandish plots and overt political agenda provided a far more explicit bias than was seen in other, non-radical news sources. It is interesting to note, therefore, the urgency with which Reynolds declares his message and his relentless avowal of its truthfulness. He says: 'Reader, this picture of [...] the aristocracy is not too highly coloured, no, nor a whit exaggerated. Ten thousand facts might be brought forward to testify its truth.'[96]

While it is clear that aristocratic males did not suffer from a fertility epidemic of the magnitude depicted in *MoCL*, their numbers *were* being reported as diminishing and doctors associated this attrition with masculine health; around the time of Reynolds's writing, an article appeared in *The Lancet* saying, 'the aristocracy of England [...] are becoming few [....] [H]ow can the unhealthy semen of such produce healthy offspring?'[97] While the accuracy of such medical opinion is debatable—at least at this point in the nineteenth century, although it would come to be much more heavily investigated in a few years' time, as we will see in the next chapter—at least some in the medical community subscribed to Reynolds's popular anti-aristocracy constructs around fertility and ability to rule. Antony Taylor writes that 'for G.W.M. Reynolds, the British aristocracy was tainted, bearing the historical stain of the Norman Conquest and carrying inherited predispositions toward tyranny'.[98] While no one who has read Reynolds could refute this claim, I posit that Reynolds's critique went a great deal further; Taylor's own use of the words 'tainted' and 'inherited predispositions' indicates his awareness of Reynolds's fascination with heredity and physiology and, by extension, the influence they had on the state of the nation. Reynolds's hostility towards noble manhood served as the perfect junction between medical assessment and middle-class Victorian values, casting suspicion not only on an aristocrat's ability to rule but on their very ability to survive.

Notes

1. G.W.M. Reynolds, *The Mysteries of the Court of London* (1848–1856). 10 vols (Boston, MA: The Oxford Society, 1920), I, p. 376; III, p. 102; III, p. 236; IV, p. 146; VI, p. 19; VI, p. 440; VI, p. 441; VII, p. 19; X, pp. 446–47.
2. Reynolds, *MoCL*, IV, p. 392.
3. Anne Humpherys and Louis James, 'Introduction', in *G.W.M. Reynolds: Nineteenth Century Fiction, Politics, and the Press*, ed. by Anne Humpherys and Louis James (Aldershot: Ashgate, 2008), pp. 1–15 (p. 6); 'Obituary', *The Bookseller*, 3 July 1879, pp. 600–01 (p. 600).
4. Anna Gasperini, *Nineteenth-Century Popular Fiction, Medicine and Anatomy: The Victorian Penny Blood and the 1832 Anatomy Act* (London: Palgrave Macmillan, 2019), p. ix.
5. James Greenwood, 'A Short Way to Newgate', *The Wilds of London* (London: Chatto and Windus, 1874), pp. 158–172 (p. 158); p. 168.

6. Diana Pérez Edelman, *Embryology and the Rise of the Gothic Novel* (London; Palgrave Macmillan, 2021).
7. I will be using The Oxford Society's privately-bound 10-volume edition of *MoCL* from 1920 and will cite references by volume and page number instead of by their original weekly publication date. It is ironic, given Reynolds's stance on the inevitable destruction of the aristocracy, that the few bound volumes of his work produced for middle- and upper-class collectors had more physical longevity than the inexpensive weekly papers produced for lower-class citizens. There are few, if any, complete and surviving collections of *MoL* or *MoCL* in newspaper form.
8. Reynolds, *MoCL*, III, p. 186.
9. Robert Saunders, 'God and the Great Reform Act: Preaching against Reform, 1831–32', *Journal of British Studies* 53:2 (2014), pp. 378–399 (p. 379).
10. Michael Diamond, *Victorian Sensation* (London: Anthem Press, 2003), p. 193.
11. Trefor Thomas, 'Rereading G.W. Reynolds' *The Mysteries of London*', *Rereading Victorian Fiction*, eds. Alice Jenkins and Juliet John (Houndmills, Basingstoke: Palgrave Macmillan, 2008), pp. 59–80, (p. 59).
12. Michael H. Shirley, 'G.W.M Reynolds, *Reynolds's Newspaper* and Popular Politics', *G.W.M Reynolds: Nineteenth Century Fiction, Politics, and the Press*, eds. Anne Humpherys and Louis James (Hampshire: Ashgate, 2008), pp. 75–89 (p. 75); Trefor Thomas, 'Introduction' to *The Mysteries of London* by G.W.M. Reynolds (1844–48), ed. Trefor Thomas (Keele: Keele University Press, 1996), pp. vii–xxiv, (pp. xv–xvii).
13. Thomas, 'Rereading', p. 59; Henry Mayhew, *London Labour and the London Poor* (London: Griffin, Bohn and Co, 1861–1862), p. 25.
14. Rohan McWilliam, 'The French Connection: G.W.M. Reynolds and the Outlaw Robert Macaire', *G.W.M. Reynolds: Nineteenth Century Fiction, Politics, and the Press*, eds. Anne Humpherys and Louis James (Hampshire: Ashgate, 2008), pp. 33–49 (p. 46).
15. Len Platt, *Aristocracies of Fiction: The Idea of Aristocracy in Late-Nineteenth-Century and Early-Twentieth Century Literary Culture* (Westport, CT: Greenwood Publishing, 2001), p. xiv.
16. Ina Zweiniger-Bargielowska, *Managing the Body: Beauty, Health, and Fitness in Britain 1880–1939* (Oxford Scholarship Online, January 2011). https://doi.org/10.1093/acprof:oso/9780199280520.001.0001.
17. David Rosen, 'The volcano and the cathedral: muscular Christianity and the origins of primal manliness', in *Muscular Christianity: Embodying the Victorian Age*, ed. by Donald E. Hall (Cambridge: Cambridge University Press, 1994), pp. 17–44 (p. 21).

18. Of George III's fifteen children, only three produced any living, legitimate offspring, not counting George IV's daughter, Princess Charlotte, who survived until adulthood only to die in childbirth.
19. Erasmus Darwin, *The Temple of Nature, Or, The Origins of Society: A Poem, with Philosophical Notes* (London: T. Bensley, 1803), N.P. ['Additional Notes IX: Hereditary Diseases'].
20. T.B. Curling, 'Observations on Sterility in Man; with Cases', *The Lancet* 82: 2079 (23 June 1863), pp. 11–13 (p. 12).
21. Anon. 'The Pathology of Genius', *The British Medical Journal*, 1.399 (20 Feb 1892), pp. 400–01 (pp. 400–01).
22. Stephen Halliday, *The Great Filth* (Stroud: History Press, 2011), p. 115.
23. Reynolds, *MoCL*, III, p. 186.
24. James Eli Adams, *Dandies and Desert Saints* (Ithaca: Cornell University Press, 1995), p. 15.
25. John Ruskin, 'Of Queens' Gardens' in *Sesame and Lilies* (1865), 12th ed. (London: George Allen, 1897), pp. 87–143 (p. 107).
26. Martin A. Danahay, *Gender at Work in Victorian Culture* (Aldershot: Ashgate, 2005) p. 2; John Potvin, *Material and Visual Cultures Beyond Male Bonding, 1870–1914: Bodies, Boundaries and Intimacy* (Aldershot: Ashgate, 2008), p. 2; Andrew Dowling, *Manliness and the Male Novelist in Victorian Literature* (Aldershot: Ashgate, 2001), p. 1; Cheryl Krasnick Warsh, 'Introduction' in *Gender, Health and Popular Culture*, ed. by Cheryl Krasnick Warsh (Ontario: Wilfrid Laurier University Press, 2011), pp. vii–xvii (p. viii).
27. Ying S. Lee, *Masculinity and the English Working Class* (Abingdon: Routledge, 2007; repr. 2013), p. 33.
28. Herbert Sussman, *Victorian Masculinities: Manhood and Masculine Poetics in Early Victorian Literature and Art* (Cambridge: Cambridge University Press, 1995), p. 11.
29. Harry Brod, 'Studying Masculinities as Superordinate Studies' in *Masculinity Studies and Feminist Theory*, ed. by Judith Kegan Gardiner (New York: Columbia University Press, 2002), pp. 161–75 (p. 162).
30. John Tosh, *Manliness and Masculinities in Nineteenth-Century Britain* (Harlow: Pearson Education Limited, 2005), p. 39.
31. Reynolds, *MoCL*, VII, p. 11.
32. Jennifer Terry and Jacqueline Urla, 'Introduction: Mapping Embodied Deviance', in *Deviant Bodies*, ed. by Jennifer Terry and Jacqueline Urla (Indianapolis: Indiana University Press, 1995), pp.1–18 (p. 1; p. 12).
33. Reynolds, *MoCL*, I, p. 175.
34. The concept of transvestism did not even appear until the early 20th century. Dr Magnus Hirschfeld, German sexologist, coined the term in his 1910 publication, *The Transvestites: The Erotic Drive to Cross-Dress* (New

York: Prometheus Books, 2003). His study was the first scientific work to conclude that the practice of cross-dressing was, in fact, divorced from the state of homosexuality. Previous to this definition, cross-dressing was viewed as a lewd and criminal act tied almost solely to the realm of male homosexual prostitution. See Vern L. Bullough's 'Transvestism: A Reexamination', *Journal of Psychology and Human Sexuality*, 4:2 (1991), pp. 53–67, (p. 53).
35. Reynolds, *MoCL*, I, p. 170.
36. Ibid., I, p. 134.
37. Reynolds, *MoL*, I, p. 7.
38. Reynolds, *MoCL*, I, p. 24.
39. Ibid., I, p. 135.
40. Ibid., I, p. 134.
41. Ibid., I, p. 135.
42. Ibid., IV, p. 205.
43. Ibid., I, p. 194.
44. Ibid., VI, p. 440.
45. Ibid., VI, p. 441.
46. Susanne Davies, 'Sexuality, Performance, and Spectatorship in Law: The Case of Gordon Lawrence, Melbourne, 1888', *Journal of the History of Sexuality*, 7:3 (January 1997), pp. 389–408 (p. 393).
47. Jennifer Terry, 'Anxious Slippage between "Us" and "Them": A Brief History of the Scientific Search for Homosexual Bodies' in *Deviant Bodies*, ed. by Jennifer Terry and Jacqueline Urla (Indianapolis: Indiana University Press, 1995), pp.129–69 (pp. 132–33).
48. Reynolds, *MoCL*, IX, p.327.
49. Ibid., III, p. 93
50. G.E.C. (George Edward Cokayne), 'Desmond', in *The Complete Peerage*, 4th ed., ed. by Vicary Gibbs, 13 vols (London: St Catherine Press, 1916), pp. 232–58 (pp. 254–58), IV.
51. Reynolds, *MoCL*, II, p. 103.
52. Ibid., I, p. 378; III, p. 93; IV, p. 458.
53. Shirley, p. 87 (italics mine).
54. Reynolds, *MoCL*, I, pp. 380–81.
55. Ibid., I, p. 382; I, p. 377.
56. Ibid., III, p. 30.
57. Dominic Janes, 'Back to the Future of the Body', in *Back to the Future of the Body*, ed. by Dominic James (Newcastle: Cambridge Scholars Publishing, 2007), pp. 1–16 (p. 7).
58. Reynolds, *MoCL*, III, p. 28.
59. Ibid., II, p. 200.
60. Ibid., V, p. 269.

61. Ibid., I, p. 134; I, p. 377; V, p. 376.
62. Christopher Hibbert, *George IV* (Harmondsworth: Penguin, 1972), p. 127; Christopher Hibbert, 'George IV (1762–1830)', in *Oxford Dictionary of National Biography*, ed. by H.C.G. Matthew and Brian Harrison (Oxford: Oxford University Press, 2004). Online ed., ed. by Lawrence Goldman (January 2008). http://www.oxforddnb.com/view/article/10541 [accessed 2 April 2023].
63. Adams, p. 21.
64. Danahay, p. 6
65. Reynolds, *MoCL*, VII, p. 142.
66. Weber, p. 13.
67. Danahay, p. 7; Adams, pp. 7–8.
68. Gwen Hyman, *Making a Man, Gentlemanly Appetites in the Nineteenth-Century British Novel* (Athens, Ohio: Ohio University Press, 2009), p. 77.
69. Reynolds, *MoCL*, I, p. 253.
70. It is unknown whether any illustrations were created of the undressed Prince in his introductory scene, so a comparison of his illustrated depiction with those of female characters shown 'in *dishabille*' [*sic*] is unable to be reached. Each volume of the Oxford Society's edition of *MoCL* contains only a single illustration in the frontispiece, though several dozen (perhaps hundred) more illustrations were published along with the text during the serial's run. Of the ten volumes in this edition and their respective ten illustrations, eight depict examples of undress or other lascivious behaviour.
71. Reynolds, *MoCL*, I, p. 159.
72. Ibid., I, p. 162.
73. Ibid., I, p. 162.
74. Ibid., VII, p. 204.
75. Ibid., II, p. 422.
76. Ibid., X, p. 434.
77. Ibid., VIII, p. 211.
78. Ibid., II, p. 189; III, p. 136; III, p. 369.
79. Ibid., I, p. 290.
80. Ibid., III, pp. 142–43.
81. Ibid., III, p. 145.
82. Hibbert, 'George IV (1762–1830)' in *Oxford DNB.*
83. Ibid., IV, p. 305.
84. Ibid., IV, p. 325.
85. Ibid., II, p. 319; II, p. 326; III, p. 56.
86. Ibid., VI, p. 19.
87. Ibid., VII, p. 328.
88. Ibid., VII, pp. 332–33.
89. Ibid., VI, p. 156; VI, p. 20.

90. Ibid., X, p. 21.
91. Ibid., VII, p. 455.
92. Though hardly an accurate biographer, Reynolds stays true to the broad strokes of history, especially as they concern his more well-known, real-life characters. While this restrains him from disciplining villainous characters with the severity Reynolds vocalises that they deserve, it also allows him to claim that his work is more truthful than it actually is—a privilege Reynolds takes advantage of many times (e.g. III, p. 145; IX, p. 434; X, p. 107 etc.).
93. Reynolds, *MoCL*, X, p. 248.
94. Ibid., III, p. 145.
95. Thomas, 'Rereading', p. 60.
96. Reynolds, *MoCL*, IX, p. 434.
97. K. Corbet, 'The Degeneration of Race', *The Lancet*, 78: 1981 (17 Aug 1861), p. 170.
98. Antony Taylor, '"Some little or Contemptible War up on her Hands": *Reynolds's Newspaper* and Empire', *G.W.M. Reynolds: Nineteenth Century Fiction, Politics, and the Press*, eds. Anne Humpherys and Louis James (Hampshire: Ashgate, 2008), pp. 99–119 (p. 105).

Works Cited

Adams, James Eli, *Dandies and Desert Saints* (Ithaca: Cornell University Press, 1995).

Anonymous, 'The Pathology of Genius', *The British Medical Journal*, 1.399 (20 Feb 1892), pp. 400–01.

Brod, Harry, 'Studying Masculinities as Superordinate Studies' in *Masculinity Studies and Feminist Theory*, ed. by Judith Kegan Gardiner (New York: Columbia University Press, 2002), pp. 161–75.

Bullough, Vern L., 'Transvestism: A Reexamination', *Journal of Psychology and Human Sexuality*, 4:2 (1991), pp. 53–67.

Corbet, K., 'The Degeneration of Race', *The Lancet*, 78: 1981 (17 Aug 1861), p. 170.

Curling, T.B., 'Observations on Sterility in Man; with Cases', *The Lancet* 82: 2079 (23 June 1863), pp. 11–13.

Danahay, Martin A., *Gender at Work in Victorian Culture* (Aldershot: Ashgate, 2005).

Darwin, Erasmus, *The Temple of Nature, or, the Origins of Society: A Poem, with Philosophical Notes* (London: T. Bensley, 1803), N.P. ['Additional Notes IX: Hereditary Diseases'].

Davies, Susanne, 'Sexuality, Performance, and Spectatorship in Law: The Case of Gordon Lawrence, Melbourne, 1888', *Journal of the History of Sexuality*, 7:3 (January 1997), pp. 389–408.

Diamond, Michael, *Victorian Sensation* (London: Anthem Press, 2003).
Dowling, Andrew, *Manliness and the Male Novelist in Victorian Literature* (Aldershot: Ashgate, 2001).
Gasperini, Anna, *Nineteenth-Century Popular Fiction, Medicine and Anatomy: The Victorian Penny Blood and the 1832 Anatomy Act* (London: Palgrave Macmillan, 2019).
G.E.C. (George Edward Cokayne), 'Desmond', in *The Complete Peerage*, 4th ed., ed. by Vicary Gibbs, 13 vols (London: St Catherine Press, 1916), pp. 232–58.
Greenwood, James, 'A Short Way to Newgate', *The Wilds of London* (London: Chatto and Windus, 1874), pp. 158–172.
Halliday, Stephen, *The Great Filth* (Stroud: History Press, 2011).
Hibbert, Christopher, *George IV* (Harmondsworth: Penguin, 1972), p. 127; Christopher Hibbert, 'George IV (1762–1830)', in *Oxford Dictionary of National Biography*, ed. by H.C.G. Matthew and Brian Harrison (Oxford: Oxford University Press, 2004). Online ed., ed. by Lawrence Goldman (January 2008). http://www.oxforddnb.com/view/article/10541 [accessed 2 April 2023].
Hirschfeld, Magnus, *The Transvestites: The Erotic Drive to Cross-Dress* (1910), (New York: Prometheus Books, 2003).
Humpherys, Ann and Louis James, 'Introduction', in *G.W.M. Reynolds: Nineteenth Century Fiction, Politics, and the Press*, ed. by Anne Humpherys and Louis James (Aldershot: Ashgate, 2008), pp 1–15.
Hyman, Gwen, *Making a Man, Gentlemanly Appetites in the Nineteenth-Century British Novel* (Athens, Ohio: Ohio University Press, 2009).
Janes, Dominic, 'Back to the Future of the Body', in *Back to the Future of the Body*, ed. by Dominic James (Newcastle: Cambridge Scholars Publishing, 2007), pp. 1–16.
Krasnick Warsh, Cherly, 'Introduction' in *Gender, Health and Popular Culture*, ed. by Cheryl Krasnick Warsh (Ontario: Wilfrid Laurier University Press, 2011), pp. vii–xvii.
Lee, Ying S., *Masculinity and the English Working Class* (Abingdon: Routledge, 2007; repr. 2013).
Mayhew, Henry, *London Labour and the London Poor* (London: Griffin, Bohn and Co, 1861–1862).
McWilliam, Rohan, 'The French Connection: G.W.M. Reynolds and the Outlaw Robert Macaire', *G.W.M. Reynolds: Nineteenth Century Fiction, Politics, and the Press*, eds. Anne Humpherys and Louis James (Hampshire: Ashgate, 2008), pp. 33–49.
'Obituary', *The Bookseller*, 3 July 1879, pp. 600–01.
Pérez Edelman, Diana, *Embryology and the Rise of the Gothic Novel* (London; Palgrave Macmillan, 2021).
Platt, Len, *Aristocracies of Fiction: The Idea of Aristocracy in Late-Nineteenth-Century and Early-Twentieth Century Literary Culture* (Westport, CT: Greenwood Publishing, 2001).

Potvin, John, *Material and Visual Cultures Beyond Male Bonding, 1870–1914: Bodies, Boundaries and Intimacy* (Aldershot: Ashgate, 2008).
Reynolds, G.W.M., *The Mysteries of the Court of London* (1848–1856). 10 vols (Boston, MA: The Oxford Society, 1920).
Rosen, David, 'The volcano and the cathedral: muscular Christianity and the origins of primal manliness', in *Muscular Christianity: Embodying the Victorian Age*, ed. by Donald E. Hall (Cambridge: Cambridge University Press, 1994), pp. 17–44.
Ruskin, John, 'Of Queens' Gardens' in *Sesame and Lilies* (1865), 12th ed. (London: George Allen, 1897), pp. 87–143.
Saunders, Robert, 'God and the Great Reform Act: Preaching against Reform, 1831–32', *Journal of British Studies* 53:2 (2014), pp. 378–399.
Shirley, Michael H., 'G.W.M Reynolds, *Reynolds's Newspaper* and Popular Politics', *G.W.M Reynolds: Nineteenth Century Fiction, Politics, and the Press*, eds. Anne Humpherys and Louis James (Hampshire: Ashgate, 2008), pp. 75–89.
Sussman, Herbert, *Victorian Masculinities: Manhood and Masculine Poetics in Early Victorian Literature and Art* (Cambridge: Cambridge University Press, 1995).
Taylor, Antony, '"Some little or Contemptible War up on her Hands": *Reynolds's Newspaper* and Empire', *G.W.M. Reynolds: Nineteenth Century Fiction, Politics, and the Press*, eds. Anne Humpherys and Louis James (Hampshire: Ashgate, 2008), pp. 99–119.
Terry, Jennifer, 'Anxious Slippage between "Us" and "Them": A Brief History of the Scientific Search for Homosexual Bodies', *Deviant Bodies*, ed. by Jennifer Terry and Jacqueline Urla (Indianapolis: Indiana University Press, 1995), pp. 129–69.
Terry, Jennifer, and Jacqueline Urla, 'Introduction: Mapping Embodied Deviance', in *Deviant Bodies*, ed. by Jennifer Terry and Jacqueline Urla (Indianapolis: Indiana University Press, 1995), pp. 1–18.
Thomas, Trefor, 'Introduction', in *The Mysteries of London* by G.W.M. Reynolds (1844–48), ed. by Trefor Thomas (Keele: Keele University Press, 1996), pp. vii–xxiv.
Thomas, Trefor, 'Rereading G.W. Reynolds' *The Mysteries of London*', in *Rereading Victorian Fiction*, ed. by Alice Jenkins and Juliet John (Houndmills, Basingstoke: Palgrave Macmillan, 2008), pp. 59–80.
Tosh, John, *Manliness and Masculinities in Nineteenth-Century Britain* (Harlow: Pearson Education Limited, 2005).
Zweiniger-Bargielowska, Ina, *Managing the Body: Beauty, Health, and Fitness in Britain 1880–1939* (Oxford Scholarship Online, January 2011). https://doi.org/10.1093/acprof:oso/9780199280520.001.0001.

CHAPTER 4

'But You Know There's a Cousin': Endogamous Marriage in Sensation Fiction

> Family likeness often has a deep sadness in it. Nature, like that great tragic dramatist, knits us together by bone and muscle, and divides us by the subtler web of our brains; blends yearning and repulsion; and ties us by our heart-strings to the beings that jar us at every movement. We hear a voice with the very cadence of our own uttering the thoughts we despise; we see eyes—ah, so like our mother's!—averted from us in cold alienation; and our last darling child startles us with the air and gestures of the sister we parted from in bitterness long years ago. The father to whom we owe our best heritage—the mechanical instinct, the keen sensibility to harmony, the unconscious skill of the modelling hand—galls us and puts us to shame by his daily errors; the long-lost mother, whose face we begin to see in the glass as our own wrinkles come, once fretted our young souls with her anxious humours and irrational persistence.[1]

Introduction

As was alluded to in the introductory chapter with Eric Hobsbawm's discussion of 'invented tradition', many customs associated with the aristocracy were primarily a nineteenth-century concoction, in which ritual and symbolic meaning emerged 'to inculcate certain values and norms of behaviour by repetition, which automatically implies continuity with a suitable historic past'.[2] Hobsbawm's argument is borne out in the genre

A. Boucher, *Science, Medicine, and Aristocratic Lineage in Victorian Popular Fiction*, Palgrave Studies in Literature, Science and Medicine, https://doi.org/10.1007/978-3-031-41141-0_4

fiction of previous chapters, where new aesthetic, cultural, and scientific ideals were integrated into aristocratic traditions as though they had always been there. For example, the faddish sensibility and modish illnesses recorded in silver fork fiction purported (with varying degrees of satire) to be the *abiding* indicator of noble heredity, rather than as a massive late-eighteenth-century paradigm shift in the performance of class and health; or how the Chartist Gothic penny-fiction of the 1840s triangulated a single recent succession crisis, the unpopularity of Hanoverian royals, and contemporary discussions about male fertility, to make claims about the biopolitical and moral constitution of the aristocracy as a historical whole.

Sensation fiction of the 1850s, '60s, and '70s, however, breaks from this pattern of invented tradition and integrates into its pages an authentically long-standing facet of aristocratic life and lineage: endogamy, or marrying within one's own group (be it social, tribal, ethnic, or familial). Talia Schaffer, in her 2016 work *Romance's Rival: Familiar Marriage in Victorian Fiction*, writes: 'Anthropology sees endogamous marriage as a null set, but the Victorian novel sees a cousin as an ideal partner.'[3] While Schaffer traces cousin-, neighbour-, and vocational marriage in literature from several genres over the course of the whole Victorian era, emphasising that familiar marriage's broad function in Victorian literature is to 'keep the woman oriented to a wider world—a local community, a family, a network of friends', this chapter takes a narrower focus: I specifically examine aristocratic endogamy in sensation fiction—a genre with massive cultural cachet for roughly fifteen years, from the late 1850s to the early 1870s, although it kept going far longer—and emphasise how depictions of these marriages engaged with genuine medical and scientific discourses around heredity and fertility (in a way that, say, Reynolds's Chartist fiction did not), as well as ideological concerns about class, race, and disability.[4] As will be seen, sensation fiction of this period uses aristocratic endogamy to understand an increasingly fluctuating socio-economic landscape and to give new registers to older concerns (and vice versa). This landscape, the paranoia of which is captured in much of sensation fiction, would eventually lead to some 'invented traditions' springing up around authentic endogamic practices (as we'll see in our discussion of the 'social season'), in an attempt to regulate access to elite circles perceived to be under threat of invasion by outsiders. Sensation fiction, feeding into the social unease which created these traditions, synthesises contemporary medical, political, gender, and class issues to capture aristocratic endogamy in its largest moment of practical and ideological transition.

I argue that sensation fiction's frequent portrayal of, and ambiguous stance toward, endogamous aristocratic marriages was a synthesis of other contemporaneous discussions about pathology, class, race, heredity, and domesticity. As will be explored fully, a burgeoning mid-century understanding of what we would now call 'genetics' (and, in a slightly later nineteenth-century development, 'eugenics'—which Angelique Richardson states was, in Britain at the end of the nineteenth century, 'primarily a discourse on class') prompted and informed a huge number of public discussions around the potential dangers or securities of marrying within or without the family circle.[5] These scientific and political discussions gained further purchase with the advent of sensation fiction and its overt interaction with race, disability, class, and familial secrecy and privacy. Sensation fiction's engagement with these topics then furthered public discourse with a reinvigorated sense of paranoia about the stability and purity of the (white, middle- and upper-class) family unit. These issues percolated through earlier texts and genres: we see eddying traces in Charlotte Brontë's *Jane Eyre* (1847), Emily Brontë's *Wuthering Heights* (1847), William Makepeace Thackeray's *Vanity Fair* (1847), Herman Melville's *Pierre* (1852), Charles Dickens's *Bleak House* (1853), Anthony Trollope's *Barchester Towers* (1857), and George Eliot's *Adam Bede* (1859) and *Silas Marner* (1861), among many others. But, as will be explored below, it is only with the rise of sensation fiction (whose generic form, content, and porousness of readerships mirrored many of these supposedly collapsing socio-economic barriers and institutional stabilities) that these anxieties were able to hit their narrative stride.

In 1863, Punch ran a semi-satirical advertisement for 'The Sensation Times, and Chronicle of Excitement', a new fictitious journal 'devoted chiefly to the following objects; namely; Harrowing the Mind, Making the Flesh Creep, Causing the Hair to Stand on End, Giving Shocks to the Nervous System, Destroying Conventional Moralities, and generally Unfitting the Public for the Prosaic Avocations of Life'.[6] The article goes on to advise every Paterfamilias to 'duly enjoy' such stories himself but to keep their contaminating influence away from his wife and children, lest the new literary genre corrupt those with a more delicate physical, mental, and moral balance.[7] *Punch*'s advertisement is able to have its cake and eat it too, highlighting the ridiculousness of both the genre's ostensible aims of producing physical sensations in its readers *and* the perceived paranoia of readers fearing the medical effects of these aims: sensation novels were seen to have 'inspired a new form of reading, one that depended first on the physical effects it inspired in the reader and secondly on the psychological effects that occurred as a result of this form of reading', that the novels 'offered the possibility of reading with the body'.[8] Beyond the alleged effects that reading thrilling stories had on the body, the advertisement's anxiously paternalistic language consolidates a number of other concerns which have since become well-analysed hallmarks of the genre: the breaking down of boundaries, the collapsing of genres and readerships, the precarious sanctity of the domestic space, the double-edged sword of surveillance, the physiological provenance of criminality and madness, the frailty of gender norms, and the blurring of class lines. That *Punch* advertises a new journal of (what was commonly dismissed as) 'kitchen fiction' to the wavering domestic authority of a (presumably) middle- or upper-class man, instead of advertising it to his servant—*and* that *Punch* also raises the question of the medical and moral fortitude of that man's own family—gets to the heart of this chapter's focus: sensation fiction's depiction of hidden pathologies and the blurring of boundaries in middle- and upper-class marriage, while also modelling the ambiguous position of the genre itself. Sensation fiction intensified concerns that nothing (not even the sanctity of family and knowability of the self) and no one (not even the most pedigreed of aristocrats) were immune from corrosion.

Sensation fiction's perceived status as a pathological and invasive genre has been well covered by critics, both consistently in the popular culture and criticism of the mid-to-late Victorian era and consistently in scholarship after the fact. The same year as *Punch*'s faux advertisement, the philosopher and ecclesiastic Henry Longueville Mansel anonymously wrote a vicious attack on sensation fiction in *Quarterly Review*, saying the novels were 'preaching to the nerves' and showed

> indications of a widespread corruption, of which they are in part both the effect and the cause: called into existence to satisfy the cravings of a diseased appetite, and contributing themselves to foster the disease, and to stimulate the want they supply.[9]

In a passage as equally hyperbolic as the ones he lambasts, Mansel goes on to say sensation fiction authors revel in their own exaggerated portrayals of the evil that comes from things like liquor, loose women, 'the peerage', and incest as severe as man 'marrying his grandmother'.[10] More than fifteen years later, John Ruskin was still echoing its sentiments (albeit about literature more generally and without Mansel's Bible-thumping or *Punch*'s tongue-in-cheek): he worries about the prevalence of 'physical corruption[,] moral disease [...] and the conditions of languidly monstrous character developed in an atmosphere of low vitality [which had] become the most valued material of modern fiction'.[11]

Perhaps these reactions aren't surprising. Sensation fiction appeared amidst rapid urbanisation, technological advancement, and significant reforms on marriage, voting rights, working-class education, and labour, and it exploited these self-same social phenomena for its own narrative ends. The genre's fixation on detective stories, which theoretically restore order and class boundaries that had been disrupted by crime and the conditions of modern life, also rely on disrupting order and class boundaries to do so: Kate Summerscale traces the mid-century rise of police forces, describing how detectives were popularly located somewhere between rational deities and prurient sneaks—their goal 'to disappear, to slip noiselessly between the classes'.[12] The genre's genesis also coincided almost perfectly with a range of other events and developments which touched certain nerves. The 1861 death of Prince Albert (one half of the married couple who redefined upper-class domesticity) created a strange punctuation mark on his contributions to class porousness not previously witnessed in the monarchy or aristocracy. Around this time was the rise of psychology and concepts of the 'double consciousness', or the fear that psychotic breaks, sleepwalking, alcohol abuse, or hypnosis could lead people to act on impulses that were a mystery even to themselves; this concept was used liberally in sensation fiction, perhaps most famously in Wilkie Collins's *The Moonstone* in which the amateur detective and narrator, Franklin Blake, realises at novel's end that he was the one who had committed the crime in a fit of drug-induced sleepwalking, leading to a further sense of self-alienation and disassociation: 'I had discovered Myself as the Thief [...] the shock inflicted on my completely suspended my thinking and feeling power. I certainly could not have known what I was about.'[13] This episode—still prescient decades later—would be repeated in one of the more sensational chapters of the 24-authored experimental novel, *The Fate of Fenella* (1892), in which a sleepwalker in aristocratic circles commits a gory murder.[14] And, certainly as important to sensation fiction and its preoccupation with the safety of knowability was Charles Darwin and his *On the Origin of Species* (1859) and *The Descent of Man* (1871), the social, scientific, and literary ripples of which will be explored much more centrally in Chaps. 5 and 6.[15] Sensation fiction was a product of the fears around liminality, modernity, and perceived social collapse and was then critiqued for the very conditions of its birth: by addressing these issues, the genre continued to stoke the fires around them.[16]

An examination of just a few individual sensation texts and their engagement with aristocratic intermarriage would be a difficult feat. It would require, firstly, too much preliminary context before getting to literary analysis; secondly, an impossible selection of a few texts from a pool of hundreds; and, thirdly, the omission of ongoing dialogues between genre and popular scientific discourse over decades. Instead, what follows is an exploration of the many scientific, historical, religious, political, and socio-economic threads which shape sensation fiction's complex stances on aristocratic endogamy and pathological heredity. I weave a wide selection of sensation fiction texts through these issues to show them in conversation and to better capture the wavering landscape and complex dynamics of the genre as a whole—much as I did with the equally large silver fork genre in Chap. 1, which shares an obsession with capturing and semi-satirically commenting on a rapidly shifting contemporary culture. Importantly, I trace how endogamy's most significant transhistorical threads (dynastic, religious, economic) took on additional registers in the Victorian era (domestic, medical heredity, and class habitus) and how those changes are borne out in the literature of the day. Further, by exploring sensation fiction (with all its reflections and refractions of *modernity* in Britain) from a more transhistorical or socio-economic vantage, I will prime my discussion of later literary genres (explored in Chaps. 4 and 5) which focus on *antiquity* in Britain, and with their much broader evolutionary and even eugenicist considerations that first get traction here.

Sensation Fiction: An Overview

Wilkie Collins's *The Woman in White* (serialised from 1859–60), followed immediately in as many years by Mrs Henry [Ellen] Wood's *East Lynne* (1861) and Mary Elizabeth Braddon's *Lady Audley's Secret* (1862) solidified sensation fiction as its own genre. This is not to say that the genre appeared fully formed, as some of the genres in other chapters do: some of the earliest examples of silver fork fiction settled into an established and never-before-seen formula almost immediately, as do the Ruritanian romances examined in the next chapter. There are many factors that go into the creation of an individual and new genre, but one of the largest is an author striking the right level of novelty at precisely the right moment in the zeitgeist. Sensation fiction, which pulled heavily from several other established genres, needed to build up to its own identity. Elements that would slowly come to be hallmarks of sensation fiction ('marital irregularities [or] a woman with a secret'; 'the excitement of gothic horror, but situated [in the] home'; 'lack of ethics and issues of power'; the replacement of 'moral certainty with moral ambiguity'; and a more theatrical or '"dramatic" method of narration') had been percolating in other fiction for more than a decade by the time Collins, Wood, and Braddon's works weaved together and heightened these elements into a more definitive mode.[17] The slow segueing of Newgate, melodramatic, Gothic, Romantic, and even realist fiction into what we now know as 'the sensation novel' makes the genre's exact genesis or the labelling of certain borderline texts difficult to pinpoint. Several sensation tropes appear, for example, in the 1847 Romantic-Gothic novels of Emily and Charlotte Brontë (*Wuthering Heights* and *Jane Eyre*, respectively) and the mid-to-late career texts of the more realist novelist Charles Dickens (Wilkie Collins's best friend and frequent collaborator). Mansel, in his rant about the godlessness and gracelessness of sensation fiction, says, 'Mr. Dickens, we regret to say, is a grievous offender in this line.'[18] In particular, Dickens's *Bleak House* (1852–53)—with its perversions of the aristocratic domestic space, its 'violated hearth', and its 'secret[s which] may take air and fire, explode, and blow up'—seemingly wrote a blueprint for the tensions around class, marriage, health, and hidden backgrounds depicted in sensation fiction for decades to follow.[19] The same year, Wilkie Collins wrote what some critics see as the first sensation novel (or its strongest, clearest precursor), *Basil* (1852)—although it was nowhere near as popular as Collins's later *The Woman in White* (1859), which is more often perceived to be the first sensation novel.[20]

The precise epicentre of sensation fiction, whether critically deemed to be *Basil*, *The Woman in White*, or some other text, doesn't particularly matter to our understanding of it. Sensation fiction is a genre with its own deeply exogamic parentage and relies heavily (and aptly for this chapter) on Wittgenstein's theory of 'family resemblances' for any attempt at a critical definition. Not only did the complexity of Victorian modernity, class issues, and medical understandings produce the many concurrent conditions needed for the creation of the genre, but the genre itself went on to inform, sensationalise, comment on, and reproduce these conditions out of which it was born. While my focus will be on sensation fiction's high period in the 1860s and '70s, the scientific and social issues discussed in the chapter (and the genre itself) extend both far before and far after this period. As such, I will include in this chapter some precursor sensation fiction from as early as the 1840s, all the way through the genre's creaky decline and diminishing returns (at least in its first, Victorian iteration) which extends as far as the 1890s. Sensation fiction is easily the largest, most porous, and most sprawling genre in this book and to treat its boundaries as firm as some of the other genres I explore would do it a disservice.

As has already been alluded, a significant trope in the sensation fiction catalogue of 'family resemblances' is the intersection of medical science, the body, and social boundaries. Commonly, sensation fiction scholars focus on the process of reading and the genre's perceived ability to influence the body. Numerous critics have worked from the argument that 'the continuity between reading and transgressive practices posed a threat to social and political stability'.[21] Andrew Mangham in his *Violent Women and Sensation Fiction* (2007) explores the reversal of this process, in which real cases of female violence, insanity, and bodily sensation were fed back into the genre's content, inspiring the very stories which was then feared would go on to produce more violence, insanity, and bodily sensation. Elizabeth Steere traces class-based views of reading with the body, with one of the perceived chief dangers being the genre's ability to blur boundaries through physical sensation: that the genre produced bodily feelings in its readers, feelings which transgressed gender, class, national, and ethnic lines and somehow tainted or Othered one's physiology through exposure to narrative. Lyn Pykett and Pamela K. Gilbert both look not only at the assumed damaging effect of sensation novels on individual bodies, but also on the public body: in her *The Nineteenth-Century Sensation Novel*, Pykett argues that sensation fiction was 'taken to be evidence of a cultural disease', while Gilbert's *Disease, Desire, and the Body in Victorian Women's Popular Novels* (1997) examines how popular fiction, and in particular sensation fiction, came to be associated with contagion and a threat to certain hegemonic identities.[22] Even sensation fiction itself commented on its own pathology: in Wilkie Collins's detective-sensation novel, *The Moonstone* (1868), the protagonist claims he is swept up by the '*detective-fever*' of finding himself embedded in the crime of a sensation plot, a term more often used by critics of this brand of popular culture.[23]

There are, of course, many other significant and growing critical discourses around sensation fiction's interrogation of medicine, science, and the body: the middle-class professionalisation of the medical field;

disability studies; mental health and burgeoning concepts of psychoanalysis; medical ethics and policy; various forms of abuse (substance, domestic etc.). But it is the genre's almost constant tension around aristocratic endogamy which perhaps most aptly reflects the sensation fiction's own fight for legitimacy and its perceived heritage as a 'mongrel' literary form of dubious generic and class provenance. In this book's focus on genre and popular fiction of the long nineteenth century, sensation fiction is the genre which perhaps is the least 'pure', which most heavily borrows from all other genres in other chapters, and whose class-based medical and scientific content is most in dialogue with its generic form.

Endogamy and Ecclesiastical Issues

Simply put, endogamy is marriage or procreation within a particular family, community, or tribe. The term is largely sociological or anthropological in use—Scottish ethnologist John Ferguson McLennan notes in his 1865 *Primitive Marriage* that

> As the words endogamy and exogamy are new, an apology must be made for employing them. Instead of endogamy we might, after some explanations, have used the word caste. But caste connotes several ideas besides that on which we desire to fix attention. On the other hand, the rule which declares the union of persons of the same blood to be incest has been hitherto unnamed, and it was convenient to give it a name. The words endogamy and exogamy (for which botanical science affords parallels) appear to be well suited to express the ideas which stood in need of names, and so we have ventured to use them.[24]

While it is unclear from his wording whether McLennan is coining these terms himself or whether he is simply one of the first to formally employ new terminology that he heard elsewhere, the terms were rarely, if ever, used in the sensation fiction which so frequently deploys them as a generic device. While fiction preceding McLennan's work (and, indeed, fiction published significantly later than it) uses other language like 'familial marriage', 'marriage of kin', 'marriage of kind', 'in-marriage', 'consanguinity', 'homogamy', or even—in the right circumstances—'incest', for the purposes of clarity I will use only 'endogamy' in this chapter, as the terms are more appropriate for the relationships sensation fiction depicts: not *necessarily* marriages between family members, but marriages between those in a closed group with a tightly shared socio-economic and ethnic background. While McLennan clearly applied 'endogamy' in racist and orientalist ways to the 'rude tribes' he studied, the term at its most basic sense can and should be applied to the aristocracy—especially given the 'primitive' (as McLennan would have it) roots and essentialised nature of class-based marriage. Endogamy is a common practice in *most* groups (especially elite ones) and is a definitive element of aristocracy itself. It is not only a way of consolidating wealth and power by keeping assets in a family or other tight circle, continuing a dynastic line, and solidifying alliances within an exclusive network, but also a means by which a group may define their membership.[25]

The more popular and less anthropological definition of endogamy (and, correspondingly, exogamy)—which is the way I am using it—is fuzzy and shifting: it is often but not always synonymous with the marriage of blood relations, and it conceptually is not the same as (but has an uneasy and mercurial overlap with) incest. Wilkie Collins's *Basil* toys heavily with this concept: Basil, the aristocratic protagonist, spends the novel trapped in a love triangle (albeit not the main love triangle the novel is known for) with his sensual, duplicitous, lower-class, and eventually diseased, secret wife Margaret, and the far more compatible, lovely, upper-class Clara, *his sister*—the latter with whom he falls into a domestic and overtly romantic (though non-sexual) relationship by novel's end. That there are seemingly no other romantic options for Basil, apart from either pathological exogamy or outright incest, presents an almost tongue-in-cheek understanding of anxieties around endogamy. A very similar pattern crops up the same year in Herman Melville's proto-sensational philosophical work, *Pierre: Or, the Ambiguities* (1852). Where Collins's vaguely incestuous love triangle was only alluded to, here Melville presents an explicitly incestuous and complex love pentagon among the upper-class Pierre, his two cousins, his illegitimate sister, and his mother—as though no other romantic options were available in either Pierre's small town or the large city to which he ventures. Melville goes so far as to rhapsodise that 'a reverential and devoted son seemed lover enough for this widow' and 'He who is sisterless, is as a bachelor before his time. For much that goes to make up the deliciousness of a wife, already lies in the sister.'[26]

The difficulty of endogamy's definition, both in the Victorian era and today, is due to significantly older ecclesiastical issues. Some relationships were deemed taboo on familial grounds despite no blood relationship between the two parties, the most famous being the prohibition of marrying a deceased sibling's spouse or a deceased spouse's sibling. Theological confusion on this matter stems from two contradictory injunctions in the Old Testament: Leviticus 18:8-18, which prohibits sexual relations with (in addition to certain blood relatives) your father's wife, your father's brother's wife, your daughter-in-law, and your brother's wife.[27] The last of these laws is directly contradicted later, in Deuteronomy 25:5-10: 'If brothers are living together and one of them dies without a son, his widow must not marry outside the family. Her husband's brother shall take her and marry her and fulfil the duty of a brother-in-law to her', or else risk punishment.[28] There is a lengthy history of debate on this issue, but this debate was reinvigorated (and had exhausted people) in the Victorian era, after the passing of The Marriage Act, 1835 ('Marriage with a Deceased Wife's Sister'), which made all subsequent marriages of a man with his deceased wife's sister automatically null and void. The Act would not be repealed until 1907.[29]

A further complicating factor was when certain expressly forbidden marriages could, under the aegis of the Catholic Church, be excused by papal or episcopal dispensation. Dispensation was even occasionally granted if the blood relationship between a bride and groom was close enough to be considered outright incest.[30] What was determined 'too close' a kinship largely depended on the vagaries of politics and the relationship between the religious official granting dispensation and the individuals requesting it. The most significant of these vagaries was seen when Henry VIII was granted papal permission to marry—and was later refused papal permission to divorce—his dead brother's widow, Katherine of Aragon, which led to England's break with the Catholic church in favour of Protestantism. This conversion would have considerable knock-on effects for endogamous aristocratic marriages in Britain and its colonies in later centuries.

The Protestant Reformation, especially in Britain, created two thorny issues which rocked marriage markets in aristocratic and royal circles. The first was the sudden lack of papal authority to (de)legitimise certain marriages based on degrees of consanguinity, and the second was an excess of children who now needed to be married off (especially girls, who could not as readily inherit or earn their own living) due to a sudden lack of convents and monasteries in which to deposit them. Protestant aristocratic families now had a far higher number of appropriate marriages to arrange (with the coinciding fear of a bad marriage dragging down the whole dynastic apparatus), and a more collective, social (and therefore a less absolute and official) decision to be made about what types of marriages were appropriate, for social, medical, economic, and political reasons. It is almost entirely due to the Protestant Reformation, Kristen Richardson argues in her 2020 social history, *The Season*, that we have two of the most enduring and identifiable of aristocratic traditions around endogamy (made increasingly dynamic in nineteenth-century literature): the 'social season' and the 'debutante'. While endogamous marriage is as old as aristocracy itself and present in communities far beyond the scope of European aristocracy, the 'Season'—Richardson argues—is one of Hobsbawm's 'invented traditions': a *ritualisation* of endogamy (steeped in false nostalgia) to help articulate the contemporary logistics and imperatives of a small ruling class, and the practical need to vet potential relationships, alliances, and newcomers.[31] In speaking of her own experience invited to be presented as a debutante, Richardson writes:

> We would lend our family name to the festivities, and I would emerge from my chrysalis, ready to carry the values I was meant to embody into the wider world, a kind of upper-class mascot. Right before I was set to go to college, to leave the privileged, interconnected world I grew up in, I was being called on to affirm that I would stay.[32]

Richard discusses at length the social and economic ideologies embedded in this practice. However, a crucial element of this exclusive ritual, which Richardson touches upon very briefly but which is felt very deeply in much mid-to-late nineteenth-century literature about aristocratic marriage, is heredity—by which I mean issues of racism and ableism. This is certainly not a chapter about the social season. In fact, despite being a genre populated with aristocrats, the social season itself is hardly touched upon in sensation fiction, despite its development being a fascinating real-life indicator of huge paradigm shifts in class and heredity practices. One of the few notable examples is in Mrs Henry Wood's 1867 *Lady Adelaide's Oath*, where the titular impoverished aristocrat only becomes a fixture of the social season once she has avoided endogamous marriage to two of her lordly cousins and instead married a rich middle-class man. This is a strange parsing of the season, in which Lady Adelaide is given access to her own class from someone outside of it, and only once she no longer needs access to the season to arrange a suitable marriage.

The development of 'the Season' does raise two issues which are pertinent to and broadly explored in sensation fiction. The first is how the long-standing norms of the aristocracy morphed in the nineteenth century to police issues of lineage, long before 'genetics' were understood. The second is how—in light of the evolving medical evidence about the marriage of close kin—sensation fiction and these new invented aristocratic traditions engaged with, challenged, repackaged, or disguised as 'romantic ritual' concerns about heredity. Beyond the socio-economic and ecclesiastical factors in endogamous marriage—which are not inconsequential—sit the medical and eugenicist ones, and indeed the former two factors might well be conceived of as placeholders for concerns that would eventually be articulated by the latter.

The Science of Endogamy: Nineteenth-Century Class Contexts

It is difficult to separate scientific concerns about endogamy and incest from ecclesiastical ones, given that most religious or social practices which try to govern elements of the natural world (like taboos around sexual relationships) are borne from, or are reconfigurations of, issues within that natural world. Empirical observations about the effects of endogamy and exogamy float through the works of religious and scientific thinkers alike and are equally rationalised by their respective disciplines. Through trial and error and via the increasing sophistication of human society, useful biological practices (the exact mechanics of which may not be understood, even if the results were) became Law.[33] That said, it was over the course of the nineteenth century that scientific research began to explain more sufficiently what before was only empirically intuited about endogamy: how traits were passed down and the degree to which repeated endogamy could be safe or dangerous. It wasn't even until the mid-eighteenth century, Maurizio Meloni argues, that ideas of heredity even began to coalesce around biology or other life sciences. Before this point, concepts 'of heredity were scattered among philosophical commentaries, encyclopedias, medical and moral treatises, and other sources'.[34] As will be discussed below, definitive evidence on genetic heredity wouldn't be discovered or popularly understood until at least the 1890s, but queries about the mechanics of inheritance were everywhere. Even in her 1859 *Notes on Nursing*, Florence Nightingale provocatively words a passage on the degenerative effects of houses—respectable, upper- and middle-class houses—being too tightly sealed and isolated, in a way that is strangely evocative of other, more anxious contemporaneous medical discourse:

> I have known cases of hospital pyæmia [blood poisoning] quite as severe in handsome private houses as in any of the worst hospitals, and from the same cause [...] it was that the uninhabited rooms were never sunned, or cleaned, or aired [...] it was that the windows were always tight shut up at night [...] you may often find *a race thus degenerating and still oftener a family*. You may see poor little feeble washed-out rags, *children of a noble stock*, suffering morally and physically, throughout their useless, degenerate lives, and yet people who are going to marry and to bring more such into the world, *will consult nothing but their own convenience as to where they are to live, or how they are to live.*[35]

Much has been made of the nineteenth-century preoccupation with social categorisation, especially in the wake of new scientific disciplines, the growth of the middle classes, and the effects of increasing European imperialism.[36] Discourse around heredity was no exception. Over the second half of the nineteenth century, scientists, scholars, reformers, and artists of all stripes grappled with human, social, and individual development and categorisation. Where did heritable characteristics come from? Which traits were dominant? Apart from dominance, which traits were 'superior'? What were the mechanics of trait transmission to offspring? Could inheritable traits be altered, manufactured, or rendered dormant in future offspring by a parent's behaviour (a theory that would come to be known as 'soft heredity', rooted in the works of French naturalist JB Lamarck)? Or was it impossible for the individual human to influence the genetic traits that they passed on (a theory that would come to be known as 'hard heredity', rooted in Charles Darwin's works)?

While it would seem that adhering to endogamic marriage would be an easy way to alleviate some of these broader anxieties around classification, the actual scientific discourse took a considerably more muddled stance. In what is now an oft-cited note to himself, a young Charles Darwin made an anxious and extended list of the pros and cons of marrying his first-cousin, Emma Wedgwood: some of the cons included a loss of time to travel and work and being 'forced to visit relatives' (presumably some of his *own*), while the benefits were that she would be 'better than a dog anyhow'.[37] What is curiously absent from his extensive list—given that he certainly would have been aware of contemporary debates around it—are concerns about their kinship proximity. Darwin would soon thereafter blame the early deaths of some of his and Emma's children on his marriage of close kin, as well as similar marriages between his and Emma's ancestors, and push very strongly for further national study of endogamy.[38] Darwin's son, George, built on this work, conducting a study of cousin-marriage (with an especial focus on aristocratic cousin-marriage) and the frequency with which their offspring ended up in asylums. The results were inconclusive, although George Darwin was willing to speculate that 'as far as insanity and idiocy go, no evil *has been shown* to accrue from consanguineous marriages'.[39] Despite his son's research, Charles Darwin notes through several of his works (with increasing frequency through the 1850s and beyond) that cross-pollination or wider variety in breeding was always preferable and led to a reduction in the transmission of harmful characteristics.[40] Darwin in particular writes that the practices within aristocratic marriage markets were a 'direct evil' for Natural Selection: in particular the significant interbreeding within a small pool, and how more than a 'moderate accumulation of wealth' interferes with the process of sexual selection.[41] Charles Darwin's own cousin, the eugenicist Francis Galton, disagreed on several fronts (although Galton had vacillating ideas around class and endogamy over his life).[42] Galton, slightly contradictorily and working from the 'soft hereditarian' stance of Lamarck's 1809 'use-inheritance' theory, wrote in 1865:

> Can we hand down anything to our children, that we have fairly won by our independent exertions? Will our children be born with more virtuous dispositions, if we ourselves have acquired virtuous habits? Or are we no more than passive transmitters of a nature we have received, and which we have no power to modify?[43]

Lamarck and Galton's 'soft hereditarian' ideas percolated through much of the nineteenth century. As we can see from the inherent contradictions in Galton's own stances—which lightly imply that one can change their own characteristics so one's children may have 'improved' hardwired characteristics—these ideas further complicate concepts of endogamy. Galton (among other burgeoning eugenicists) believed firmly in careful selection of mates from among similar groups, while simultaneously signalling the potential for 'improvement'. He sought betterment of groups through the conscious selection of certain traits to forward through breeding (or eradicate through being bred out) *and* the improvement of the individual through moral, social, and intellectual gains which he presumed could be passed down through soft heredity. This of course only muddied the waters of categorisation further if one was recommended to marry 'in group' to maintain the supposedly inherent traits of that group, while simultaneously being told they could 'improve' their social, moral, intellectual, and genetic standing. The social barriers put around endogamic practices were often at odds with the aspirational and even utopian eugenicist practices which often went hand-in-hand with them. These incongruous stances were necessary to the emergence and effect of sensation fiction. By its nature, sensation fiction presents the *promise* of a return to order, a promise which it rarely delivers in full: for the genre to exist in the first place, it must either disrupt that order or—worse still—show the supposed order never existed at all.

The ideas that Galton espouses, however paradoxical, were popular; the prolific sensation novelist Ouida echoes (if not works directly from) his contradictions in her essay 'The Sins of Society'. In soft hereditarian terms, she lambasts aristocrats for ignoble behaviour which could damage the purity of their line and encourages them to return to high standards of taste and comportment. But much more severely does she blame the disappearance of 'high breeding' on exogamy, with its 'continual alliance of old families with new wealth'.[44] So harshly does she categorise the wealthy bourgeois as hereditary 'interlopers' that they are barely considered human: they infiltrate the aristocracy 'only to borrow and bask in [it] as pigs in the mud'.[45] Ouida's Galtonian evocation of bestiality trickles through her fiction, too: in her late sensation novel *The Massarenes* (1897), her narrator disdains and fears 'a new-born plutocracy creeping upward on its swollen belly like the serpent of scripture'—a sentiment that had been and would continue to be expressed on class, race, and disability.[46]

Though somewhat anticipated but not fully covered in the high period of sensation fiction, these 'high' vs 'soft' hereditarian tensions came to a head in 1893, with a major public debate between English biologist Herbert Spencer and German evolutionary biologist August Weismann in *Contemporary Review* in which they

> effectively put on trial perhaps the era's most widely-held belief about biological inheritance: that an organism—specifically a human one—could acquire throughout its individual life beneficial biological features which could in turn be passed on to the next generation. The debates also touched on a deep concern for late Victorians, self-determination, which in this context would involve the ability of the individual to control the biological destiny of future generations.[47]

These debates were one of many watershed moments discussing heredity in the public arena. Spencer—the coiner of the term 'survival of the fittest' and disparagingly called a 'Neo-Lamarckist'—agreed in large part with Galton about soft heredity and yet also the need for endogamous, eugenicist breeding. He argued that 'mixed' offspring would not fare as well as those bred from within the same groups:

> Concerning parents of pure and mixed breeds respectively, severally tending to reproduce their own structures in progeny, we may therefore say, figuratively, that the house divided against itself cannot withstand the house of which the members are in concern

and that the traits from two different groups 'would be forever fighting for dominance, or some traits would simply be needless or even burdensome'.[48] Mary Elizabeth Braddon characterises the mixed-class heroine of her 1863 *Aurora Floyd* in such oscillating terms, although for a more generally positive result. Aurora, the daughter of an upper-class banker and a provincial 'second-rate' actress, often has her physical traits read as though they were vying for prominence:

> A phrenologist would have told you that [Aurora's] head was a noble one [...] Miss Floyd knew very little about her poor mother's history [...] The county families accepted and made much of the rich banker's heiress; but they were not slow to say that Aurora was her mother's own daughter, and had the taint of the play-acting and horse-riding, the spangles and the sawdust, strong in her nature.[49]

Neither Aurora's nobility nor her earthiness of spirit is able to fully win out in the end, each battling the other in Spencer's soft hereditarian terms, although without sharing Spencer's agony over such a prospect.

Weismann, on the other hand, was categorised as a hard hereditarian and 'ultra-Darwinist' and shared none of Spencer's worry about 'mixed breeds', which put him at odds with 'much of the Victorian biological community'.[50] Weismann, since at least the 1880s, discussed the natural dilution of traits as generations go on, while still making space for the odd tenacious trait that could endure through, or crop up again after, several generations—which was of course a major concern for sensation fiction. '[M]y father was not to be deceived', writes Wilkie Collins in his novella *Mad Monkton* (published in his 1887 collection *Little Novels*), 'He knew where the hereditary taint still lurked; he viewed with horror the bare possibility of its reappearing one day in the children.'[51] The upper-crust Monktons of Wincot Abbey are insular family (the parents are cousins) with persistent strain of inherited madness. As such, they

> bore a sad character for want of sociability in our country. They never went to other people's houses, and [...] never received anybody under their own roof. Proud as they all certainly were, it was not pride, but dread, which kept them thus apart from their neighbors [*sic*]. The family had suffered for generations past from the horrible affliction of hereditary insanity, and the members of it shrank from exposing their calamity to others.[52]

Beginning strongly in the hard hereditarian camp, Collins then hedges his bets considerably. Young Alfred Monkton, 'the best, the kindest, the sanest of human beings' seems to have reduced his chances of madness through 'study and retirement', edging into soft hereditarian territory.[53] He then lapses back into madness when researching the potential moral failings of an ancestor who *went away from England* and died abroad; Alfred's obsession with finding and returning his uncle's body to the endogamous family plot for burial becomes a crux of the text, and Collins worries over the blurred lines of genetic inheritance, behaviour, and the more complex and liminal constructions of mental health—if the *fear* of madness is what drives Alfred mad.

Sir Arthur Conan Doyle is more definitive in his portrayal of the horrors of 'a constitutional and hereditary taint', partially born from exogamy, in his 1894 sentimental sensation story 'The Third Generation'.[54] The story opens with a medical examination of a young baronet, Sir Francis, and his doctor's confusion as Sir Francis swears to his own chastity despite his syphilitic symptoms. A brief jaunt through the family tree traces the syphilis back to Sir Francis's grandfather, a man with a

> sinister reputation [as] a notorious buck of the [eighteen] thirties, who had gambled and duelled and steeped himself in drink and debauchery until even the vile set with whom he consorted had shrunk away from him in horror, and left him to a sinister old age with the barmaid wife whom in some drunken frolic he had espoused.[55]

Here Doyle edges the text far more into the hard hereditarian camp. On the one hand, the text lauds the moral improvement of subsequent generations and their tendency to avoid exogamous barmaids as either wives or casual companions. But the text ultimately views these moral improvements as hereditarily futile in the face of genetics and congenital inheritance: Sir Francis's father's 'innocent life did not save him', nor, by extension, did it save Sir Francis: 'If I were heir to my grandfather's sins as well as to their results I could understand it, but I am of my father's type [...] The coarse and animal is abhorrent to me.'[56] Sir Francis, due to be married the following week and knowing that no amount of virtue on his own part can change the realities of congenital inheritance, chooses to commit noble suicide and opt out of the vagaries of hard heredity all together.

While these medical and scientific debates raged on without a full understanding of the mechanisms behind them, research on chromosomes began. Beginning with the Silesian friar Gregor Mendel's pea plant breeding experiments in the 1850s and '60s (and the coining of the terms 'recessive' and 'dominant' to describe inherited traits), research continued through the 1860s with the work of German biologist Walther Flemming, which gained traction in the 1880s. The chromosome theory of inheritance became widely accepted in the early 1900s via the research of American geneticist Walter Sutton and German cytologist Theodor Boveri, who independently published research at the same time which identified chromosomes as the transmitters of genetic information and thereby explains how Mendel's theories of inheritance (which had been somewhat forgotten and was rediscovered in 1900 by Huge de Vries) work. This theory, called alternatively the Boveri-Sutton or the Sutton-Boveri chromosome theory (based on who is believed to have published their work first) remains at the centre of our understanding of genetics to date, and put paid to the simultaneous 'soft' vs. 'hard' hereditarian debates, firmly in favour of 'hard' hereditarianism.

The resolution of the several threads seen here, of course, could not have been reasonably anticipated by sensation authors, beyond the incomplete observations they may have witnessed Darwin, Galton, Spenser, and others writing in the 1850s and '60s and subsequently integrated into their own works. Though issues of genetic inheritance and endogamic marriage were imperfectly understood at this time, one can feel their slow boil, especially where social and economic issues overlapped. In her 2008 *Family Likeness: Sex, Marriage, and Incest from Jane Austen to Virginia Woolf*, Mary Jean Corbett persuasively argues, in the light of the complexities provided by the intersection of several (often racist and classist) scientific disciplines, that if 'the *biological* wisdom of cousin-marriage [...] was increasingly debated after 1860 regarding its effects on the offspring of such unions, then the *social* wisdom of this form of kin-marriage among elites was rarely in doubt'.[57] That these vying discourses were deemed to be 'up in the air' meant there was space for their exact configuration to be contested in literature and culture. Sensation fiction, which was catalysed roughly in tandem with greater research on heredity, takes an understandably ambiguous stance—at least initially, when the logistics were imperfectly understood.

We have already seen in Wilkie Collins's *Basil* the 'rock-and-a-hard-place' mentality of marrying either endogamously or exogamously, but few authors were more ambiguous about aristocratic marriage than the prolific sensation novelist Mrs Henry Wood. Her *East Lynne* (1861), one of the 'big three' sensation novels, is overtly obsessed with the socio-economic, health, and lineage impacts of exogamous marriage and yet strangely abstains from—or perhaps straddles the line between—soft and hard hereditarian stances, as well as endogamous vs. exogamous relationships. Wood shows the misery of her heroine, Lady Isabel Vane, a vulnerable but healthy woman from a sickly aristocratic line who enters a kindly but (on her end) transactional marriage to the middle-class Archibald Carlyle, which reduces her physically and emotionally. Her unhappy affair with a man from her own social circle results in her severe disability when she flees her marital home and is presumed dead in a train crash. Archibald remarries, while the exceedingly disfigured Isabel is sneakily able to return to her former home disguised as a governess to help raise her own children. Both her exogamous and endogamous relationships produce misery, mistrust, and children who die young. Although deeply maudlin and moralistic, the message that Lady Isabel should have been content in her exogamous marriage is not borne out by the novel's dynamics.

Wood's *Lady Adelaide's Oath*, which was briefly examined earlier, has a similar dynamic, whose ambiguity and hereditarian confusion is expressed slightly in reverse from *East Lynne*: the impoverished Lady Adelaide is unhappily engaged to one of her aristocratic cousins but in love with another. When her fiancé is presumably killed—at the hands of her lover, she suspects—she becomes revolted by her remaining endogamous option and mercenarily marries a wealthy middle-class man before practically disappearing in a fug of depression from the entire second half of the text, of which she is the titular heroine. Strangely, that she produces a copious number of children in her exogamous marriage seems irrelevant, as her offspring are all inconsequential and uninteresting to her and are barely alluded to in the text.

Wood's ambiguity continued through her whole career: her novella *The Surgeon's Daughters* (appearing in her 1887 collection of short stories, *Lady Grace and Other Stories*) is perhaps her strangest text about aristocratic marriage, made all the stranger considering the huge scientific developments on understandings of heredity which had been made in the intervening decades. Wood was seemingly unconvinced. Her novella is about the aristocratic Florence Erskine, whose 'ancestors had been the highest of the high [...] descended originally from royalty'.[58] Her uncompromising father, 'Gentleman Erskine' deems no one a good enough romantic candidate for his daughter and she lives in a state of near seclusion. In a series of increasingly ridiculous and punitive events, she enacts a prophecy by going on an ill-advised donkey ride with her secret bourgeois lover and is immediately struck down and killed by lightning.[59] There is no real marriage or lineage potential at all in *The Surgeon's Daughters*; her options are either a slow obliteration in an endogamic but non-sexual relationship with her father or a quick obliteration in an exogamic one. It may be easy to see that the pattern emerging for Wood, as well as for many others, was that there was no future whatsoever for the aristocracy. But if one considers the aristocracy to often be a stand-in for greater concerns about health and science, a high-profile body as a canary in the coal mine, then the ambiguous, anxious stance of Wood and other sensation novelists makes more sense as they approach the fin de siècle, despite increasing scientific clarity on genetics.

As will be expanded in the next two chapters and their interest in human origin, to look backwards is also a way of looking forward. One can better understand the future by sifting through the past. Although pedigree was, of course, always a concern of the aristocracy, genealogical interest started in earnest as middle-class people constructed family trees, which then took on an additional role 'as a means to trace the appearance of disease in families'.[60] Collins, in his *The Woman in White* interrogates ideas of middle- and upper-class obsessions with health heredity through Frederick Fairlie and Lady Laura (nee Fairlie) Glyde, a physically delicate uncle and niece. Of course, as curiosity about tracing disease and 'latent' heritable characteristics reached a particular frenzy in the 1860s, latent 'madness', criminality, or other vaguely defined 'moral diseases' appear as staple plot points of the sensation genre. This is where Collins becomes particularly playful: as Lady Glyde becomes more unravelled, her shared *physical* frailty with her uncle is replaced by more dire comparisons to other *mentally* frail relatives on the same side of the family: the ruthless brainwashed obsessive, Madame Fosco, and the tremulously, gently mad Anne Catherick. Just as Collins fully insinuates that Laura's pathological heredity cannot be escaped, triangulated as it is among so many ill relatives, Collins then uncasts the die: Laura is perfectly sane and healthy and has merely been gaslit into believing otherwise by her psychotic husband, Sir Percival, in order to cover up the stain of his *own* heredity, which he dies attempting to cover up. But familial connection is not given a full reprieve when one remembers that Laura's marriage to Sir Percival was endogamic: he was a close friend and peer of her father (a closer familial link than all of Laura's other ill relatives). Sir Percival is insistent, albeit for his own shady financial gain, that he and Laura never procreate, and she goes on to marry again after his death—exogamously, happily, with at least one child. A sense of pathology *does* seem to emanate from Laura's predecessors, manifesting itself in novel ways in the family circle; Laura, it seems, is the outlier or has only made a narrow escape through her exogamous marriage.

All of the 'big three' sensation novels—*East Lynne, The Woman in White*, and *Lady Audley's Secret*—unravel a mystery around medical inheritance, making pathological lineage an immediate and almost necessary feature of sensation fiction from its inception. In Braddon's *Lady Audley's Secret*, Lady Audley's strange and startling revelation is not that she is even lower class than her aristocratic husband, Sir Michael, suspected, nor that she's already (and now bigamously) married, but rather that she is 'A MAD WOMAN!', a woman with 'latent insanity! Insanity which might never appear; or which might appear only once or twice in a lifetime. It would be a *dementia* in its worst phase, perhaps; acute mania [...] she has the hereditary taint in her blood.'[61] These are some of the more notable and oft-cited passages in all of sensation fiction. So rattled are the novel's other characters by this revelation that they remain in an aggressively endogamic circle, marrying and socialising only with those from within a very tight, verifiable frame of reference. Lady Audley is hyperbolically banished to a mental asylum on the Continent where she is reported to have rather conveniently died, presumably restoring safety and order. Hidden predispositions—like the vicious, criminal madness of the girlish, seemingly innocent Lady Audley—are often (but not always) portrayed as the result of exogamic deceit: usually from class or racial miscegenation hidden from at least one of the parties in the marriage. Lady Audley manages a particularly skilful sleight of hand to infiltrate the aristocracy: beginning as the married and abandoned lower-class Helen Talboys, she reinvents herself as Lucy Graham, the kindly, practical, modest governess of a respectable family in Sir Michael's social circle. Verified in her new, performative identity, she secures a marriage proposal from Sir Michael—an exogamic marriage, to be sure, but a less egregious one than she would have had as Helen Talboys. Now as Lady Audley, she slowly reinvents herself again, but this remodelling happens directly before the eyes of Sir Michael, his family, and the entire neighbourhood who notice nothing amiss: where she was once a shy, modest educator, she now becomes a vapid, childish coquette, conforming more fully to how she believes an aristocratic woman should be read. 'She hated reading or study of any kind, and loved society [...] loll[ing] on one of the sofas in her luxurious dressing-room, discussing a new costume for some coming dinner party [...] with her jewel box beside her, upon the satin cushions, and Sir Michael's presents spread out in her lap.'[62]

That our antiheroine morphs in full view of the text's characters, segueing from one perfect alignment with a class to another, without them saying a word about her overt transformation, speaks to the characters' acceptance of 'soft heredity'. Lady Audley's transformation is a seemingly organic, positive class betterment at a nearly genetic level, the improvements of which would no doubt pass down to her children with Sir Michael. But Braddon takes a far different scientific stance than her characters: the novel is a lecture in hard heredity. Nothing Lady Audley does can change her lower-class, criminalistic, mad roots and, in fact, it's the other way around: it is her 'subpar' duplicitous origins that change her behaviour, not her behaviour that changes her essence. Even the doctor, in his diagnosis, states that she 'has the cunning of madness, with the prudence of intelligence. I will tell you what she is [...] She is dangerous!'[63] If that weren't enough, she and Sir Michael are unable to have children together, despite their respective proven fertilities (each has a child from a previous marriage) and their clearly sexual relationship which spans several years (in addition to Sir Michael lurking outside her bedchamber, Braddon hints at their mutual participation in a sexual fetish, where Sir Michael treats his wife like a prized daughter and Lady Audley dresses like a little girl).[64] While hard hereditarianism doesn't necessarily insinuate in its own right that people of different groups cannot or should not mix, the novel falls clearly on the side of hard hereditarian endogamy: while characters grow and change *within* their own class group, one's latent medical or class-based characteristics cannot evolve out of it through behaviour.

Sensational Endogamy: A Question of Race

Much of the literature seen so far in this chapter has dealt with the threat of an exogamous spouse introducing disability, criminality, and vice into an aristocratic bloodline, despite previous vetting. William Thackeray's shrewd liar (and potential murderess) Becky Sharp, Braddon's mad bigamist Lady Audley, Wood's unchaste and severely disfigured Lady Isabel Vane, or even Louisa May Alcott's ruthless wig-and-fake-teeth-wearing actress Jean Muir in *Behind a Mask* (1866), among just a few, all possess enough cultural cache to pass as governesses in middle- and upper-class homes without drawing much suspicion to their histories and true natures.[65] After segueing into domestic familiarity, they then manage to falsify enough greater class *habitus* to marry into the aristocracy (except Lady Isabel Vane, who passes as middle class to *descend* the social ladder). But even the most extreme of these examples are—one their surface—about lower-middle-class white women. Marrying more exogamously than that is often unfathomable in sensation fiction, and perhaps for good *formulaic* reasons (if narratively lazy ones.)

Firstly, sensation fiction relies upon surprise, upon the façade of safety being suddenly and shockingly torn down. If an interloper is too clearly coded to either readers or other characters as a disruptor, then both narrative tension and narrative plausibility are lost. The secrets of a disruptor must remain easily obfuscated from view: a 'functioning' disability, an easily concealed criminal past, a class origin only a step or two lower than their spouse's, a sense of racial uniformity. If a secret cannot be concealed from the beginning, it cannot shock later. Pamela K. Gilbert argues in her work, 'Sensation fiction and the medical context' (2016) that realist fiction appropriated 'many of the [new] techniques of the clinical medical gaze of the period', in which a clinician was now 'an expert who relied on vision and non-verbal cues to interpret the condition of the patient, of which the sufferer might not even be aware'.[66] While Gilbert goes on to trace sensation fiction's indebtedness to this realist mode, I argue that to some not-insignificant extent, sensation fiction undoes the objective gaze of both realist fiction and clinicians by seeking to obscure from reader and medical experts alike the truth of a character's being—at least partially and at least initially.

Secondly, the aristocracy is heavily idealised in fiction and popular culture as white. The reality of European royal and aristocratic groups marrying those from non-white backgrounds for millennia is often not borne out in their media representations. The implication of 'whiteness' is especially ridiculous considering that definitions of whiteness have changed over the centuries and have, at various points and from various perspectives, excluded the Irish, Slavic groups, Jewish groups, and Mediterranean groups, among others—groups that have all have clear historical records of marrying into the British aristocracy. Concepts of racial 'purity', whatever the historical reality, are prevalent in narratives of British aristocracy, as they reinforce pedigree in other ways: an aristocrat's line has always been white, has always been upper class, has always been British, has always been able-bodied. In sensation fiction, the aristocracy (or any other authoritative group about to be infiltrated) must be portrayed as totally white to enhance the sense—and shock—of reverse colonisation. As I argue below, sensation fiction most overtly portrays the fears of a lower-class invasion, which then extends to a disabled and criminal invasion, which itself is often thinly veiled discourse on racial invasion of authoritative white spaces.

Finally, the reality of many of the texts examined in this monograph—no matter their overt political stances—is that they ultimately glamorise and flatter aristocrats. Even the silly and performative silver fork aristocrats or the viciously useless Chartist fiction ones are often afforded high levels of beauty, (sinister) appeal, and narrative centrality. This continues in sensation fiction, where aristocrats are often duped, but understandably so, by cunning individuals who have fooled everyone else. There is a certain compassion expressed in this genre towards aristocrats who have their long, pure lineage challenged and broken down by a secretive outsider or by the rumblings of modernity. Even in the genre's most critical and grotesque examples of nobility, there is a certain amount of pity expressed in the idea that aristocrats can no longer operate in this world. Formally, this is somewhat a necessity: the novels would scarcely evoke 'sensation' if they did not first reinforce a norm they ultimately seek to smash (the importance of hierarchy, the safety of whiteness). Exogamous difference therefore cannot be too visible, lest it render the aristocratic dupe into a figure of ridicule and disbelief rather than one of at least tepid sympathy.

That is not to say that sensation fiction never deals with unhidden disability, unhidden lower-class origins, or unhidden racial difference. It does, and often. But *unhidden* differences rarely make it into sensation fiction's portrayals of aristocratic marriage. What is more interesting, beyond the genre's fear of marrying your (white) cousin or fear of marrying a (white, slightly lower-class) stranger, is how the genre formally uses these discussions of class, disability, and familial proximity to more quietly integrate and address issues of race. Much like its ostensibly middle-class governesses who wormed their way into the aristocracy through slow acclimatisation, sensation fiction uses its more *familiar*, more *understood* discourses on class and disability to slip in a deeper, subtler discourse on race.

Given the overlaps among imperial, anthropological, and social reform thought during this period, it would be difficult to have a book about nineteenth-century class and lineage not also be, somewhat, a book about race. This is especially true in a chapter about nineteenth-century discourses on endogamy which, as we have seen, was frequently portrayed as a safe, emotionally mature option for upper-class white British people while simultaneously being evidence of savagery if performed by McLennan's 'rude tribes'. Alfred Henry Huth's pseudo-biological work *The Marriage of Near Kin* (1875) is one of the clearer examples of this double standard. He 'enumerated the evils of both race-mixture with and incest among "primitive" peoples [while] he simultaneously advocated in-marriage among white elites'.[67] Ann Laura Stoler, in her work on race, colonialism, and sex, expounds on this dynamic, tracing the conceptual link of sexual 'subversion to perversion, racial purity to conjugal *white endogamy*, and thus colonial politics to the management of sex.[68]

The overlap in thought between nineteenth-century social reformers and anthropologists, both of whom studied sexual behaviour, meant that—before long—the white British lower classes were being discussed much in the same imperial, animalistic terms as non-white, non-British people.[69] In Reverend Andrew Mearns's 1883 pamphlet, *The Bitter Cry of Outcast London*, he frames the godlessness and incest prevalent in London slums as direct threat to 'the Union'—that is, the United Kingdom and empire—and compares the way those in slums live to that of 'a wild beast [...] these pestilential human rookeries [...] which call to mind what we have heard of the middle passage of the slave ship'; they must learn to 'live as something better than the uncleanest of brute beasts. This must be done before the Christian missionary can have much chance with them.'[70] From there, it is an easy link to conceptualise the infiltration of a lower-class person into a middle- or upper-class family in terms of miscegenation. As imperialism, anthropology, and social reform grew, Adam Kuper writes in his study on incest and literature, so did the scope of novelists' fascination with 'the incest taboo and its origins', subtly trickling up the social ladder, dissecting 'family dynamics, the pathologies of domestic authority, the dangerous intimacy of brother-sister relationships, and the love affairs between cousins'.[71]

Sensation novelists' integration of race and empire into these discussions is perhaps not surprising: the sudden fluidity between class groups or between gender roles or between city and country—thematic staples of the genre—could be matched on a global scale: it was easy to go abroad to the far-flung reaches of empire, to gain or lose wealth and status, to absorb different mores and behaviours, to radically modify oneself. Such is the case with George Eliot's sensational fable, *Brother Jacob* (1864) in which an ambitious young confectioner's assistant, David Faux, robs his family so he can escape to the West Indies and turn his modest sugar-working skills into some sort of imperial sugar fortune. Through 'a general idea of America as a country where the population was chiefly black, it appeared to him the most propitious destination for an emigrant who [...] had the broad and easily recognizable merit of whiteness'.[72] His plan abroad fails and he returns to England to a new town, under a new name, with new gender and racial dynamics. Now the owner of a cake and pastry business whose 'place in the scale of rank had not been distinctly ascertained', he enters the realm of the feminine by stepping on the toes of local housewives, who only slowly admit the superiority of his professionally manufactured desserts.[73] The highly symbolic sugar in the text is more than a way for David to segue between gender norms or class groups as his business grows: it serves as a sort of racial, moral, and mental contaminate which filters through all walks of English society. Once described as having a 'pasty visage [and a] lipless mouth', David comes back from the West Indies with a 'sallow complexion', as a man of ambiguous provenance who, to put it in the most clearly racialised terms, 'could charm the ears of Grimsworth Desdemonas'.[74] When he is finally adopted into the community and even manages a courtship of a local high-ranking blonde English rose, Penny Palfrey—despite it being 'not a good sign when men looked so much above themselves for a wife'—his infiltration is undone by a surprise encounter with his jovial, mentally disabled brother, Jacob, whose relentless pursuit of sweets uncovers David's hidden identity.[75] Despite Jacob's congeniality and guilelessness, Eliot still leans on the shared lineage of the two men and pairs Jacob's disability with David's criminal deceit and racial tainting, indicating a corruption of the blood that predates and compounds David's overseas exposure. That David and Jacob's last name is 'Faux' is telling: it is unclear whether it is pronounced in the French way to indicate a falsehood, or anglicised as 'Fox', to heighten both David's slyness and an animalistic (and therefore racialised) reading. Either way implies a sort of foreign exogamy—David, and by extension his disabled brother, are not quite English or not quite human. David is ousted from Grimsworth, Penny has a narrow marital escape, and social, gender, racial, and economic order are restored to the community.

Of all the sensation novelists, it is probably Wilkie Collins who has the most overt—as well as the most varied—engagement with race and lineage. One of the protagonists of his1866 novel *Armadale*, Ozias Midwinter, lives in terror that his biracial and criminal heritage will lead him—as though in a prophecy or fated by blood—to harm another man.[76] Collating the research of Natalie Schroeder, Jenny Bourne Taylor, and Peter Thoms, literary scholar Monica M. Young-Zook cites Midwinter's inherited imperial trauma and its effects on his body, that his 'interpellation of his father's guilt [...] turns him into a colonial subject who carries the guilt of the violent crimes of the colonizer'.[77] Collins's aptly named 1869 play, *Black and White: A Love Story in Three Acts*, is set in Trinidad in the 1830s and centres around the Count de Layrac whose courtship with an heiress is broken upon the discovery that the Count has non-white ancestry and is therefore legally a slave. Joshua Gooch, in his work on *Black and White*, cites Collins's works with other mixed-race characters, such as in *Miss or Mrs?* (1873) and *The Guilty River* (1886), whose 'imagined biological division represented their interior psychological conflicts'.[78] Even Collins's Miserrimus Dexter in *The Law and the Lady* (1875), who gets far more critical attention as a severely disabled, aristocratic, 'grotesque horror', is given a racially and endogamously charged undercurrent.[79] Born without legs, descriptions of Dexter often veer into the animalistic as a 'terrible creature', a 'new Centaur, half man, half chair', with 'the ears of a dog', who—when not in his wheelchair, upon which he sits 'like a bird alighting on its perch'—walks 'the floor as lightly as a monkey, on his hands', while his upper-class status and his extreme, cultivated male beauty of the ancient Greek variety present him as a character at odds within himself.[80] Peter J Kitson in his work on colonialism and race in literature extensively traces the eighteenth- and nineteenth-century understandings of racial 'gradation' in terms of medical, zoological, and anthropological study, citing many contemporaneous theories and studies out of which Dexter seemed to be born.[81] While Dexter's mental and physical disabilities take centre stage in both the text and its criticism, it is his orientalisation and seedily endogamous living arrangements which add a richer dimension to the portrayals of those disabilities—especially because Dexter's romantic advances are repeatedly rejected by middle-class white women. He lives with, and is tended to very intimately by, his mentally disabled cousin Ariel, 'an idiot', 'a mere vegetable', 'an imperfectly developed animal in shapeless form'.[82] Collins makes explicit the non-white reference of Ariel's name, beyond its connection to Shakespeare's exotic sprite, saying that she speaks with a voice more like 'Caliban's'.[83] Though both Dexter and Ariel are clearly stated to be white, the subtext is anything but: their house is perpetually dark with elements of the 'Oriental' and the 'barbaric', and the squalid domestic living arrangements of the two disabled cousins (despite their upper-class status) evokes the ethnic anthropology of McLennan and anticipates the socio-ethnic reform work of Mearns.[84]

It is Collins's 1869 detective novel *The Moonstone*, however, which has perhaps his most complex, subtle, and anxious dynamics around race, aristocracy, and endogamy. As such, *The Moonstone* can stand in as a case study for many of the other dynamics in sensation fiction which often more quietly portray aristocratic endogamy as an issue of racial provenance in addition to the more overt concerns about fertility, disability, and class. Racial heredity is winked at in *The Moonstone's* very premise, though never explicitly stated in Collins's work: the titular Moonstone enters the English house of Lady Verinder as the sacrilegious spoils of British imperialism in India by Colonel Herncastle. With the story highlighting the clear criminality of Colonel Herncastle from its first chapter (whether his criminality is implicit, or somehow a contaminate picked up through his exposure to foreign lands), he then presents his aristocratic descendants with an even more malicious inheritance than the potential for latent vice and corruption. Loathing his family and knowing the physical dangers associated with possessing the Moonstone, Herncastle leaves it in his will to his aristocratic niece Rachel Verinder in the hopes that the Brahmin priests dedicated to the stone's recovery will hunt her down and harm her in the process: 'here was our quiet English house suddenly invaded by a devilish Indian Diamond—bringing after it a conspiracy of living rogues, set loose on us by the vengeance of a dead man.'[85] Of all the potential colonies or spheres of foreign influence to choose, Collins was deliberate in selecting India, as Sue Lonoff writes in her *Wilkie Collins and his Victorian Readers* (1982): not only did Collins begin his work barely a decade after the bloody and much publicised Indian Rebellion of 1857 which 'conjured up grim visions in most Victorian minds', but Collins's use of Brahmins enabled a greater reflection and refraction on the close-knit aristocratic community in which he sets his story.[86] The group of elite Brahmins that Collins depicts are so single-minded and so insular that their ultimate success (in recovering the Moonstone and restoring their cultural purity and hegemony) would lead to their ultimate destruction and demise. They

> forfeited their caste in the service of the god. The god had commanded that their purification should be the purification by pilgrimage. On that night, [they] were to part [...] Never more were they to look on each other's faces. Never more were they to rest on their wanderings, from the day which witnessed their separation, to the day which witnessed their death.[87]

The endogamous overlaps between the upper classes in Britain and India was, by this point, already being considered in medical and public arenas: an 1873 article in *The Lancet*, authored by a British Surgeon in India, praises Galton and cites what he considers to be successful inbreeding practices within Indian high castes: The Brahmins 'have intermarried only amongst themselves for centuries, and the result is a race of men who have a great degree secured for themselves [...] the attributes [...] of governing. [A Brahmin in a crowd] is as distinctive as would be a racer amongst a mixed troop of ordinary hacks, hunters, and carriage-horses'. He urges *The Lancet*'s editor to reconsider the journal's stances against inbreeding—the goal, the author states, is for the British to do the same lest they 'succumb before this race'.[88]

In a racially tinged and very actively pernicious form of heredity, the question becomes if the Verinder line is already tainted—either through their ancestor's natural criminal depravity or from him 'mixing' with non-white foreigners—or, if not yet tainted, if an exogamous danger is inevitable. Influences from both inside and outside the family unit are presented as equally unsafe, which compounds the major romantic subplot: Rachel's difficult decision to marry one of her cousins, Franklin Blake or Godfrey Ablewhite (whose ludicrous name connotes safety and Britishness). When the Moonstone is stolen from Lady Verinder's house, one of the cousins, Franklin Blake, takes up the mantel of detective. Here the secure knowability of the family group and even the self is again put to the test: Rachel acts both suspicious herself *and* as though she suspects Franklin (himself recently home from abroad and perhaps likewise under an exogamous racial influence) of wrongdoing. She severs their romantic relationship and leans towards a cousin-marriage with the more starkly white and British Godfrey Ablewhite. A long detective story ensues and its tensions about race relations are sophisticated, if never clearly resolved.

The fear of non-white invasion in an aristocratic home takes on a particularly intimate tone when all its occupants have their rooms and persons searched after the supposed Indian robbery; and, indeed, the whole crime pivots around the mystery of missing bedclothes. A prurient register reached, the tracking down of the supposed Brahmin thieves then leads to a shock of the familiar: the aristocratic family lawyer meets with a 'rather remarkable-looking', but presumably white man, so 'dark in complexion that we all set him down in the office for an Indian, or something of that sort'; the man is, indeed, one of the Brahmins, who affects white British respectability so well that his skin colour merely confuses the lawyer's clerks.[89] He is 'carefully dressed in European costume', speaks 'in an excellent selection of English words', and dazzles the lawyer by being 'the perfect model of a client [...] he did what none of my own countrymen had ever done, in all my experience of them—he respected my time'.[90] The Brahmin's rationality, reasonableness, and Britishness of manner far exceed that of many of the white British characters in the text (as the lawyer immediately compares: 'Mr Luker was, in every respect, such an inferior creature to the Indian'), made all the more remarkable considering the Brahmin cannot adequately imitate other Indians of a lower caste: in an attempt to distract and amuse the white aristocrats earlier in the novel, they give a 'very bad and clumsy imitation' of lower-caste jugglers.[91] This dynamic leads the reader down several indecisive paths with regard to Collins's meditations on endogamy. The Brahmin can undermine the conceptual stranglehold that 'whiteness' has in defining 'Britishness'—that what is central to national identity is *not* racial in nature. Or, conversely, the Brahmin's ability to merge seamlessly into British mores is the result of a Britain already stained by proximity to other cultures: he can assimilate because there is no 'pure' Briton to whom he might be contrasted. Or, even more sinisterly, if this scene is to be read as a *performance* of Britishness, then British identity may be appropriated and taken away—that national identity (if not, in this case, racial identity) may be performed and manipulated by the Brahmin as easily as any of sensation fiction's lower-middle-class governesses who become aristocrats. The relative safety felt by the white British lawyer (and, presumably, reader) at the Brahmin's familiarity is undercut by the Brahmins' actual murderous capabilities and intentions: they *do* eventually recover the Moonstone through violence. Of course, this sense of danger is undercut again, or perhaps reified, when several other white British characters (including the paragon of British masculinity, Godfrey Ablewhite) are revealed in the text to be just as dangerous, if not more so, than the Brahmins. The dual plot threads of non-white threat and aristocratic cousin-marriage twine around each other, making a slight mockery of Rachel's 'informed' decision to marry one of her family members. Despite its detective novel format and reliance on logic and clarity to restore order, Collins shows that the security of knowledge and truth can only extend so far.

Other racially liminal characters populate the text and bring their own complications. The local doctor's assistant, Mr Ezra Jennings is described in non-white terms, with 'gipsy darkness' and features 'found among the ancient people of the East, so seldom visible among the newer races of the West'.[92] A mixed-race metaphor is literalised in this 'freak of Nature': he has 'piebald hair' with perfectly black and perfectly white patches co-existing and even zigzagging into each other, but never blending.[93] To compound his duality, to increase a sense of oriental decadence and taboo, and also perhaps to mitigate a sense of violence and danger, Jennings says he was 'born with [a] female constitution'.[94] Despite his ties to the community and his increasingly respectable profession, 'nobody knew much about him in our parts [...] and, right or wrong, we none of us liked him or trusted him'.[95] A depressive and opium user, Jennings is the one who ultimately solves the mystery. Jennings is foiled, or perhaps complemented, by a guest of the family, happily taken into their company when Jennings is not:

> [T]he celebrated Indian traveller, Mr Murthwaite, who, at risk of his life, had penetrated in disguise where no European had ever set foot before. This was a long, lean, wiry, brown, silent man [...] It was rumoured that he was tired of the humdrum life among the people in our parts, and longed to go back and wander off the tramp again in the wild places of the East.[96]

As opposed to the objectively far less dangerous Jennings who is an established, if exotic, member of the community, Murthwaite is trusted and lauded for his infiltration of an exogamous group (because it's infiltration in the 'correct' direction) *and* for his ambiguous racial appearance (because its ultimate provenance is white British and its dubious character enables white British imperialism). Despite his supposed racial and national liminality and sensitivity towards the Brahmins' cause, he still advises the family to chop up, and thereby destroy the religious importance of, the Moonstone, rather than simply repatriating it—though such sacrilegious destruction may well lead to Rachel's death anyway through the Brahmins' vengeance.[97] The negotiation of ambiguous race in the text is difficult, with an inverse relationship between their propensity towards harm versus the respect they're granted; although it is Jennings who moderately wins out with his far greater narrative centrality and influence on the plot. Formally, Jennings becomes more endogamous than Murthwaite: Jennings's is allowed to narrate an important late segment of the novel, from his journal extracts, while Murthwaite is never given narrative reins. Despite order being restored by the novel's end with the exogamous realms separating back into their own spheres and a corresponding reduction of danger, the novel itself, with Jennings's narration, is a product of racial exogamy and is far less stable than it appears. It blends the old-fashioned epistolary novel and—by the late 1860s—the now more established sensation fiction with newer genres, the detective story and the Imperial Gothic, and undermines all of them in doing so. As discussed earlier in this chapter, the primary narrator, Franklin Blake, is discovered to have been the initial thief of the jewel, having been drugged and taking it during a sort of sleepwalking blackout. His actions, witnessed by his uncomprehending cousin Rachel, were then hidden from him during his tenure as amateur detective. Not only is knowledge within the family unit flawed and hidden, but so is the reliable knowledge of the self. The fear of the double consciousness or the split self, the fear that 'several selves could inhabit one body', grew in prominence around this time with the rise of psychological study, criminology, and especially the sensationalism of crime in the press, but it also tapped into deeper issues around heredity and what secrets could lurk in the blood, hidden even from the self, and what influences could trigger its exposure.[98]

Three years later, Sheridan Le Fanu would take this fear to its most extreme conclusion in his sensation-horror short story, 'Green Tea' (1872), in which a respectable man—probably not coincidentally also named Jennings—drinks green tea (a slightly tongue-in-cheek reference to sensation fiction's more frequently feared Chinese opium) and triggers what amounts to an Orientalist psychotic break. Jennings repeatedly envisions a demonic monkey that blasphemes and encourages him to commit crimes, tapping into a side of himself he did not recognise.[99] That *The Moonstone* requires other narrators than Franklin Blake—narrators who span class and racial groups, including Ezra Jennings—illustrates Blake's own lack of cohesion and authority, as well as a lack of narrative 'purity': the 'truth' of the case, subjective though that is, cannot be fully known by an individual or even an endogamous group. For better or worse, exogamy is here to stay.

Conclusion

Due to many factors (including an oversaturated market, new and more experimental forms of literature, the death of some of its better authors, and the increasing preachiness of those remaining), sensation fiction burned out in popularity during the 1870s, although that would not stop it from continuing as a genre until its rebirth in the present day. The genre made a huge resurgence in the 2010s through its rebranding as the contemporary domestic thriller which, in terms of form and content, is still modelled almost exactly on the 1850s and 1860s version—the domestic secrets of a wealthy white elite, class tensions heightened by exogamous marriages, narrative conventions centred around new technology and media, and a not-very-updated fascination with criminality and madness. Instead of latent insanity, we focused on the figure of the psychopath, or the personality disorder, or the trauma reaction: Lady Audley, seen through a modern psychotherapy lens.

Of course, in the popular mindset the aristocracy has since made way for new types of elite groups, but the demographic remains depressingly endogamous—hence why modern sensation fiction is almost exclusively about wealthy white women. And of course, as cousin-marriage, even to the third and fourth degree, has hugely fallen out of popularity in white western groups with the rise of the understanding of genetics, this is no longer a specific concern of the genre. Even by the end of the nineteenth century, only shortly after these debates had settled, sexologist Havelock Ellis would see endogamic marriage as flawed in a completely different way, with heredity not even considered. Ellis articulates that those who have grown up together or obtained a certain degree of familiarity would 'dull the sensual attraction' and rob a relationship 'of its power to call forth the necessary erethistic excitement required to produce sexual tumescence'.[100] But that is not to say that modern sensation fiction has not retained the fears of soft- and hard-inherited traits. Texts like Emma Donoghue's *Room* (2010), Liane Moriarty's *Big Little Lies* (2014), Paula Hawkins's *The Girl on the Train* (2015), and Gillian Flynn's *Gone Girl* (2012), among many others, overtly query if children can inherit the vices and failings of a parent and, if so, how—through learned behaviour or hardwired, through their genetic code?

Despite conclusive genetic research making it into the public consciousness throughout the 1890s and 1900s, artistic licence would still be taken with ideas of heredity. In fact, the preoccupation of inheritance and its influence on—and the influence it receives *from*—modernity would spur itself on in novel and grander ways as the century drew to a close. Darwin's ideas, which here had been percolating through sensation fiction, would be co-opted or conspicuously ignored in the literature of the 1880s, 1890s, and 1900s for far greater and more certain ideological purposes. In the following two interconnected chapters, I explore the anxiety around heredity and the aristocratic body far beyond a mere few generations and focus on how authors manifested aristocracy at the beginnings of humanity itself.

Notes

1. George Eliot, *Adam Bede* (1859) (London: Penguin, 2008), p. 46.
2. Eric Hobsbawm, 'Introduction: Inventing Traditions', *The Invention of Tradition*, eds. Eric Hobsbawm and Terence Ranger (Cambridge: Cambridge University Press, 1983; repr. 2020), pp. 1–14 (p. 1).
3. Talia Schaffer, *Romance's Rival: Familiar Marriage in Victorian Fiction* (Oxford: Oxford University Press, 2016), p. 38.
4. Ibid., p. 199.
5. Angelique Richardson, *Love and Eugenics in the Late Nineteenth Century: Rational Reproduction and the New Woman* (Oxford: Oxford University Press, 2003), p. xvii; Richardson goes on to say that eugenics in Germany was more primarily associated with mental health and in the United States it was used in relation to race.
6. Anonymous, 'Prospectus of a New Journal', *Punch*, Saturday, 9 May 1863) p. 193 (p. 193).
7. Ibid.
8. Laurie Garrison, *Science, Sexuality and Sensation Novels* (Houndmills, Basingstoke: Palgrave Macmillan, 2011), pp. xii–xiii.
9. Anonymous [Henry Mansel], 'Sensation Novels', *Quarterly Review* 113: 506 (April 1863), pp. 481–514, (pp. 482–83).
10. Ibid., p. 488.
11. John Ruskin, 'Fiction—Fair and Foul' (1880) in *The Complete Works of John Ruskin, LL.D.*, vol 16 of 26 (New York: Bryan, Taylor & Company, 1894), pp. 153–219 (pp. 155–56).

12. Kate Summerscale, *The Suspicions of Mr Whicher, or the Murder at Road Hill House* (2008) (London: Bloomsbury, 2009), p. 52.
13. Wilkie Collins, *The Moonstone* (1868) (London: Penguin 1986), p. 359.
14. 'Rita' [Eliza Humphreys], 'Chapter 7' in *The Fate of Fenella* (1892) (Tunbridge Wells, Solis Press, 2016), pp. 35–39.
15. Susan D. Bernstein makes the connection between Darwin and sensation fiction explicit in the opening sentence of her article: 'On 24 November 1859 Charles Darwin's *On the Origin of Species by Means of Natural Selection* was published; two days later the first instalment of Wilkie Collins's *The Woman in White* appeared'. See 'Ape Anxiety: Sensation Fiction, Evolution and the Genre Question', *Journal of Victorian Culture* 6:2 (2001), pp. 250–71 (p. 250).
16. Raymond Williams, *The Country and the City* (London: Chatto and Windus, 1973)—who applied these anxieties to mid-century literature more generally; Winifred Hughes, *The Maniac in the Cellar: Sensation Novels of the 1860a* (Princeton: Princeton UP, 1980), pp. 59–60.; Michael Diamond, *Victorian Sensation* (London: Anthem Press, 2003), pp. 189–90.
17. Lyn Pykett, *Wilkie Collins* (Oxford: OUP, 2005) p. 16; Judith Flanders, *The Invention of Murder* (London: Harper Press, 2011); Anna Gasperini, *Nineteenth Century Popular Fiction, Medicine, and Anatomy: The Victorian Penny Blood and the 1832 Anatomy Act* (Houndmills: Palgrave Macmillan, 2019); Hughes, *Maniac*, p. ix; p. 24.
18. Anonymous [Mansel], p. 488.
19. Hughes, *Maniac*, p. ix; Charles Dickens, *Bleak House* (1853) (Ware: Wordsworth Classics, 2001) p. 304.
20. Diamond, *Victorian Sensation*, p.195; Patrick Brantlinger alludes to *Basil*'s early sensation theatrical leanings in 'What is "Sensational" About the "Sensation Novel"?', *Nineteenth-Century Fiction* 37:1 (Jun., 1982), pp. 1–28 (p. 5); Pykett, *Wilkie Collins*, p. 91; The *Westminster Review* contemporaneously called it as much when it was first published and pre-empted the exact wording and criticism that would come nearly a decade later, by saying that though *Basil* took its 'incidents from real life, [it seems] to revel in scenes of fury and passion, such as, happily, real life seldom affords' and keeps men in 'low passions': Anon., 'Progress of Fiction as an Art', *Westminster Review Vol. LIX and LX, January and April 1853*, American Edition (New York: Leonard Scott & Co, 1853), pp. 179–96 (p. 195).
21. Barbara Leckie, *Culture and Adultery: The Novel, the Newspaper, and the Law, 1857–1914* (Philadelphia: University of Pennsylvania Press, 1999), p. 21.
22. Pykett, *Sensation Novel*, p. 7.

23. Collins, *The* Moonstone, p. 160.
24. John F. McLennan [M'Lennan], *Primitive Marriage: An Inquiry into the Origin of the form of Capture in Marriage Ceremonies* (Edinburgh: Adam and Charles Black, 1865), pp. 48–49, footnote.
25. McLennan, *Primitive Marriage*, p. 47. Adam Kuper, *Incest and Influence: The Private Life of Bourgeois England* (Cambridge, MA: Harvard University Press, 2009), pp. 23.; Kristen Richardson, *The Season: A Social History of the Debutante* (New York: Norton, 2020), p.14.
26. Herman Melville, *Pierre, or The Ambiguities* (1852) (London: Penguin, 1996), p. 5; p. 7.
27. *The Holy Bible*, Authorised King James Version (online), Leviticus 18:8-18.
28. *The Holy Bible*, Deuteronomy 25:5-10.
29. United Kingdom, 'The Marriage Act 1835', 5&6 Will.4 c.54 (1907): 'All marriage which shall hereafter be celebrated between persons within the [ecclesiastically] prohibited degrees of consanguinity or affinity shall be absolutely null and void to all intents and purposes whatsoever'; United Kingdom, 'The Deceased Wife's Sister's Marriage Act 1907', 7 Edw.7 c.47 (1907).
30. Joanna of Naples required such permission to make her marriage to her nephew Ferdinand II of Naples legal, as did second cousins Isabella I of Castile and Ferdinand II of Aragon.
31. Kristen Richardson, *The Season*, p. 2.
32. Ibid., p. 6.
33. See Joseph Henrich's *The Secret of Our Success: How Culture is Driving Human Evolution, Domesticating our Species, and Making us Smarter* (Princeton: Princeton University Press, 2015) for several sociological and historical examples.
34. Maurizio Meloni, 'Nineteenth Century: From heredity to Hard Heredity', *Political Biology* (2016), pp. 32–63 (pp. 37–38).
35. Florence Nightingale, *Notes on Nursing* (1859), 1st American edition (New York, D. Appleton and Company, 1860), pp. 31–32; italics mine.
36. See Edward Beasley's *The Victorian Reinvention of Race: New Racisms and the Problem of Grouping in the Human Sciences* (New York: Routledge, 2010); Catherine Hall, Keith McClelland, and Jane Rendall, *Defining the Victorian Nation; Class, Race, Gender and the Reform Act of 1867* (Cambridge: Cambridge University Press, 2000); Damon Ieremia Salesa, *Racial Crossings: Race, Intermarriage, and the Victorian British Empire* (Oxford: Oxford University Press, 2012).
37. Charles Darwin, 'Second note [July 1838]', *Darwin Correspondence Project*. University of Cambridge library archives: DAR 210.8:2, accessed 25 July 2022, available: https://www.darwinproject.ac.uk/tags/about-darwin/family-life/darwin-marriage#.

38. Luka Fenner, Matthias Egger, Sebastien Gagneux, 'Annie Darwin's death, the evolution of tuberculosis and the need for systems epidemiology', *International Journal of Epidemiology* 38:6 (Dec 2009), 1425–28.
39. George H. Darwin's 'Marriages between first cousins in England and their effects.' *Journal of the Statistical Society* XXXVIII Part II (2): 1875, pp. 153–184 (p. 168), italics original.
40. Robert Ornduff, 'Darwin's Botany', *Taxon* 33:1 (Feb 1984), pp. 39–47.
41. Charles Darwin, *Descent of Man, and Selection in Relation to Sex*, vol 1 (London: John Murray, 1871) p. 163; pp. 169–70.
42. David Burbridge, 'Francis Galton on twins, heredity and social class', *The British Journal for the History of Science*, 34:3 (2001), pp. 323–340 (p. 335): Galton suggests that the aristocracy are particularly prone to insanity due to their endogamic practices, while elsewhere blaming exogamic aristocratic practices on noble lines dying out, due to marrying rich heiresses—their heiress status and lack of brothers implying family fertility problems.
43. Francis Galton, 'Hereditary Talent and Character, Second Paper', *Macmillan's Magazine* 12 (1865), pp. 318–27 (pp. 321–22).
44. Ouida, 'The Sins of Society', *Views and Opinions* (London: Methuen & Co., 1895), pp. 1–33 (p. 30).
45. Ibid., p. 3; p. 5.
46. Ouida, *The Massarenes* (New York: R.F. Fenno & Company, 1897), p. 125.
47. Paul J. Niemeyer, 'The Royal Red-Headed Variant: "The prisoner of Zenda" and the 1893 Heredity Debates', *College Literature* 42:1 (Winter 2015), pp. 112–138, p. 113.
48. Herbert Spencer, 'The Inadequacy of "Natural Selection"', *Contemporary Review* 63:153–66 (1893), pp. 439–56 (p. 450); Niemeyer, p. 119.
49. Mary Elizabeth Braddon, *Aurora Floyd* (1863), ed. P.D. Edwards (Oxford: Oxford University Press, 1996), p. 11; pp. 20–21.
50. Niemeyer, p. 119.
51. Willie Collins, *Mad Monkton* (1887) (Moscow: It's Time to Read Original, ОФОРМлеНИе [Forming], 2018), p. 7.
52. Ibid., p. 5.
53. Ibid., p. 9.
54. Sir Arthur Conan Doyle, 'The Third Generation', *Round the Red Lamp: Being Facts and Fancies of Medical Life* (London: Methuen & Co, 1894), pp. 46–64 (p. 54).
55. Ibid., p. 55.
56. Ibid., p. 56.
57. Mary Jean Corbett, *Family Likeness: Sex, Marriage, and Incest from Jane Austen to Virginia Woolf* (Ithaca, NY: Cornell University Press), p. 13 (italics mine).
58. Mrs Henry (Ellen Price) Wood, *The Surgeon's Daughters* in *Lady Grace and Other Stories*, 3 vols (London: Richard Bentley & Son 1887), III, pp. 45–167 (pp. 72–73).

59. For a more robust discussion of class, gender, the gaze, and the role of lineage in these three Wood texts, see my article, 'Incorporeal and Inspected: Aristocratic Female Bodies and the Gaze in the Works of Mrs Henry Wood', *Women's Writing*, 28:1 (2021), pp. 90–106.
60. Niemeyer, p. 15.
61. Mary Elizbeth Braddon, *Lady Audley's Secret* (1862) (London: Penguin, 1998), p. 340; p. 372.
62. Ibid., p. 55.
63. Ibid., p. 372.
64. Braddon writes that Lady Audley dresses 'like a child tricked out for a masquerade [...] as girlish as if she had just left the nursery', p. 55.
65. One could, of course, never include the cartoonishly French, low-bred, drunken, and almost criminally morbid Madame de la Rougierre from Sheridan Le Fanu's from his Gothic-Sensation novel *Uncle Silas* (1864), as she fails to fool anyone.
66. Pamela K. Gilbert, 'Sensation fiction and the medical context', *The Cambridge Companion to Sensation Fiction*, ed. Andrew Mangham (Cambridge: Cambridge University Press, 2013), pp. 182–95 (p. 183).
67. Corbett, *Family Likeness*, p. 13; Huth, himself from a patrician merchant banking family with aristocratic ties, married his own first cousin. Their nearly thirty-year marriage produced no children.
68. Ann Laura Stoler, *Race and the Education of Desire: Foucault's* History of Sexuality *and the Colonial Order of Things* (Durham: Duke University Press 1995), p. 184 (emphasis mine).
69. Corbett, *Family Likeness*, p. 8.
70. Andrew Mearns, *The Bitter Cry of Outcast London: AN Inquiry into the Condition of the Abject Poor* (London: James Clarkes & Co, 1883), p. 5; p. 8; pp. 24–25.
71. Kuper, *Incest and Influence*, pp. 27–28.
72. George Eliot, 'Brother Jacob' (1864) in *The Lifted Veil* and *Brother Jacob*, ed. Helen Small (Oxford: Oxford University Press, 1999), pp. 49–87 (p. 51).
73. Ibid., p. 51; p. 61; p. 67.
74. Ibid., p. 51.
75. Ibid., p. 69.
76. Midwinter's mother is said to be a 'woman of the mixed blood of the European and the African race with the northern delicacy in the shape of her face and the Southern richness in its colour'. See Wilkie Collins, *Armadale* (1864) (London: The Electric Book Company, 2001), p. 33.
77. Monica M. Young-Zook, 'Wilkie Collins's Gwilt-y Conscience: Gender and Colonialism in *Armadale*', *Victorian Sensations: Essays on a Scandalous Genre*, eds. Kimberly Harrison and Richard Fontina (Colombus: Ohio State University Press, 2006), pp. 234–45 (p. 236).

78. Joshua Gooch, '"Damn all White Men and Down with Labor": Race and Genre in Wilkie Collins & Charles Fechter's *Black and White*', *Nineteenth Century Theatre and Film*, 49:1 (2002), pp. 71–87, p. 72.
79. Wilkie Collins, *The Law and the Lady* (1875) (London: The Electric Book Company, 2001), p. 252.
80. Ibid., p. 252; p. 251; p. 253; p. 255; p. 252.
81. Peter J. Kitson, *Romantic Literature, Race, and Colonial Encounter* (Houndmills: Palgrave, 2007).
82. Collins, *The Law and the Lady*, p. 255.
83. Ibid., p. 254.
84. Ibid., p. 259; p. 265.
85. Collins, *The Moonstone*, p. 67.
86. Sue Lonoff, *Wilkie Collins and his Victorian Readers: A Study in the Rhetoric of Authorship* (New York: AMS Press, 1982), p. 223.
87. Collins, *The Moonstone*, pp. 525–26.
88. M.C. Furnell, 'Hereditary Improvement', *The Lancet* 102:2602 (July 12, 1873), pp. 62–63.
89. Collins, *The Moonstone*, p. 324.
90. Ibid., pp. 324–25.
91. Ibid., p. 327; 107.
92. Ibid., p. 371
93. Ibid., p. 371.
94. Ibid., p. 422.
95. Ibid., p. 187.
96. Ibid., p. 101.
97. Ibid., p. 109.
98. Summerscale, *Suspicions*, p. 137.
99. Jennings's last words in his suicide note are 'I therefore write, but I fear very confused, very incoherently. I am so *interrupted*, disturbed' (emphasis mine). Sheridan Le Fanu, 'Green Tea', *In a Glass Darkly* (1872) (London: Eveleigh Nash & Grayson LTD, 1923), pp. 1–51 (p. 46).
100. Eliot, *Adam Bede*, p. 46.

Works Cited

Anonymous, 'Progress of Fiction as an Art', *Westminster Review Vol. LIX and LX, January and April 1853*, American Edition (New York: Leonard Scott & Co, 1853), pp. 179–96.

Anonymous, 'Prospectus of a New Journal', *Punch*, Saturday, 9 May 1863, p. 193.

Anonymous [Henry Mansel], 'Sensation Novels', *Quarterly Review* 113: 506 (April 1863), pp.481–514.

Beasley, Edward, *The Victorian Reinvention of Race: New Racisms and the Problem of Grouping in the Human Sciences* (New York: Routledge, 2010).
Bernstein, Susan D., 'Ape Anxiety: Sensation Fiction, Evolution and the Genre Question', *Journal of Victorian Culture* 6:2 (2001), pp. 250–71.
Boucher, Abigail, 'Incorporeal and Inspected: Aristocratic Female Bodies and the Gaze in the Works of Mrs Henry Wood', *Women's Writing*, 28:1 (2021), pp. 90–106.
Braddon, Mary Elizabeth, *Aurora Floyd* (1863), ed. P.D. Edwards (Oxford: Oxford University Press, 1996).
Braddon, Mary Elizabeth. *Lady Audley's Secret* (1862) (London: Penguin, 1998).
Brantlinger, Patrick, 'What is "Sensational" About the "Sensation Novel"?', *Nineteenth-Century Fiction* 37:1 (Jun., 1982), pp. 1–28.
Burbridge, David, 'Francis Galton on twins, heredity and social class', *The British Journal for the History of Science,* 34:3 (2001), pp. 323–340.
Collins, Wilkie, *Armadale* (1864) (London: The Electric Book Company, 2001a).
Collins, Wilkie. *The Law and the Lady* (1875) (London: The Electric Book Company, 2001b).
Collins, Wilkie. *Mad Monkton* (1887) (Moscow: It's Time to Read Original, ОФОРМлеНИе [Forming], 2018).
Collins, Wilkie. *The Moonstone* (1868) (London: Penguin 1986).
Conan Doyle, Sir Arthur, 'The Third Generation', *Round the Red Lamp: Being Facts and Fancies of Medical Life* (London: Methuen & Co, 1894), pp. 46–64.
Corbett, Mary Jean, *Family Likeness: Sex, Marriage, and Incest from Jane Austen to Virginia Woolf* (Ithaca, NY: Cornell University Press).
Darwin, Charles, *Descent of Man, and Selection in Relation to Sex*, vol 1 (London: John Murray, 1871).
Darwin, Charles. 'Second note [July 1838]', *Darwin Correspondence Project.* University of Cambridge library archives: DAR 210.8:2, accessed 25 July 2022, available: https://www.darwinproject.ac.uk/tags/about-darwin/family-life/darwin-marriage#.
Darwin, George H., 'Marriages between first cousins in England and their effects.' *Journal of the Statistical Society* XXXVIII Part II (2): 1875, pp. 153–184.
Diamond, Michael, *Victorian Sensation* (London: Anthem Press, 2003).
Dickens, Charles, *Bleak House* (1853) (Ware: Wordsworth Classics, 2001).
Eliot, George, *Adam Bede* (1859) (London: Penguin, 2008).
Eliot, George. 'Brother Jacob' (1864) in *The Lifted Veil* and *Brother Jacob*, ed. Helen Small (Oxford: Oxford University Press, 1999), pp. 49–87.
Fenner, Luke, Matthias Egger, Sebastien Gagneux, 'Annie Darwin's death, the evolution of tuberculosis and the need for systems epidemiology', *International Journal of Epidemiology* 38:6 (Dec 2009), 1425–28.
Flanders, Judith, *The Invention of Murder* (London: Harper Press, 2011).
Francis Galton, 'Hereditary Talent and Character, Second Paper', *Macmillan's Magazine* 12 (1865), pp. 318–27.

Furnell, M.C., 'Hereditary Improvement', *The Lancet* 102:2602 (12 July 1873), pp. 62–63.
Garrison, Laurie, *Science, Sexuality and Sensation Novels* (Houndmills, Basingstoke: Palgrave Macmillan, 2011).
Gasperini, Anna, *Nineteenth Century Popular Fiction, Medicine, and Anatomy: The Victorian Penny Blood and the 1832 Anatomy Act* (Houndmills: Palgrave Macmillan, 2019).
Gilbert, Pamela K., 'Sensation fiction and the medical context', *The Cambridge Companion to Sensation Fiction*, ed. Andrew Mangham (Cambridge: Cambridge University Press, 2013), pp. 182–95.
Gooch, Joshua, '"Damn all White Men and Down with Labor": Race and Genre in Wilkie Collins & Charles Fechter's *Black and White*', *Nineteenth Century Theatre and Film*, 49:1 (2002), pp. 71–87.
Hall, Catherine, Keith McClelland, and Jane Rendall, *Defining the Victorian Nation; Class, Race, Gender and the Reform Act of 1867* (Cambridge: Cambridge University Press, 2000).
Henrich, Joseph, *The Secret of Our Success: How Culture is Driving Human Evolution, Domesticating our Species, and Making us Smarter* (Princeton: Princeton University Press, 2015).
Hobsbawm, Eric, 'Introduction: Inventing Traditions', *The Invention of Tradition*, eds. Eric Hobsbawm and Terence Ranger (Cambridge: Cambridge University Press, 1983; repr. 2020), pp 1–14.
The Holy Bible, Authorised King James Version (online).
Hughes, Winifred, *The Maniac in the Cellar: Sensation Novels of the 1860a* (Princeton: Princeton University Press, 1980).
Kitson, Peter J., *Romantic Literature, Race, and Colonial Encounter* (Houndmills: Palgrave, 2007).
Kuper, Adam, *Incest and Influence: The Private Life of Bourgeois England* (Cambridge, MA: Harvard University Press, 2009).
Leckie, Barbara, *Culture and Adultery: The Novel, the Newspaper, and the Law, 1857–1914* (Philadelphia: University of Pennsylvania Press, 1999).
Le Fanu, Sheridan, 'Green Tea', *In a Glass Darkly* (1872) (London: Eveleigh Nash & Grayson LTD, 1923), pp 1–51.
Lonoff, Sue, *Wilkie Collins and his Victorian Readers: A Study in the Rhetoric of Authorship* (New York: AMS Press, 1982).
McLennan [M'Lennan], John F., *Primitive Marriage: An Inquiry into the Origin of the form of Capture in Marriage Ceremonies* (Edinburgh: Adam and Charles Black, 1865).
Mearns, Andrew, *The Bitter Cry of Outcast London: AN Inquiry into the Condition of the Abject Poor* (London: James Clarkes & Co, 1883).
Melville, Herman, *Pierre, or The Ambiguities* (1852) (London: Penguin, 1996).
Meloni, Maurizio, 'Nineteenth Century: From heredity to Hard Heredity', *Political Biology* (2016), pp. 32–63.

Nightingale, Florence, *Notes on Nursing* (1859), 1st American edition (New York, D. Appleton and Company, 1860).

Niemeyer, Paul J., 'The Royal Red-Headed Variant: "The prisoner of Zenda" and the 1893 Heredity Debates', *College Literature* 42:1 (Winter 2015), pp. 112–138.

Ornduff, Robert, 'Darwin's Botany', *Taxon* 33:1 (Feb 1984), pp. 39–47.

Ouida, *The Massarenes* (New York: R.F. Fenno & Company, 1897).

Ouida. 'The Sins of Society', *Views and Opinions* (London: Methuen & Co., 1895), pp. 1–33.

Pykett, Lyn, *Wilkie Collins* (Oxford: OUP, 2005).

Richardson, Angelique, *Love and Eugenics in the Late Nineteenth Century: Rational Reproduction and the New Woman* (Oxford: Oxford University Press, 2003).

Richardson, Kristen, *The Season: A Social History of the Debutante* (New York: Norton, 2020).

'Rita' [Eliza Humphreys], 'Chapter 7' in *The Fate of Fenella* (1892) (Tunbridge Wells, Solis Press, 2016).

Ruskin, John, 'Fiction—Fair and Foul' (1880) in *The Complete Works of John Ruskin, LL.D.*, vol 16 of 26 (New York: Bryan, Taylor & Company, 1894), p. 153–219.

Salesa, Damon Ieremia, *Racial Crossings: Race, Intermarriage, and the Victorian British Empire* (Oxford: Oxford University Press, 2012).

Schaffer, Talia, *Romance's Rival: Familiar Marriage in Victorian Fiction* (Oxford: Oxford University Press, 2016).

Spencer, Herbert, 'The Inadequacy of "Natural Selection"', *Contemporary Review* 63:153–66 (1893), pp. 439–56.

Stoler, Ann Laura, *Race and the Education of Desire: Foucault's* History of Sexuality *and the Colonial Order of Things* (Durham: Duke University Press 1995).

Summerscale, Kate, *The Suspicions of Mr Whicher, or the Murder at Road Hill House* (2008) (London: Bloomsbury, 2009).

United Kingdom, 'The Deceased Wife's Sister's Marriage Act 1907', 7 Edw.7 c.47 (1907a).

United Kingdom, 'The Marriage Act 1835', 5&6 Will.4 c.54 (1907b).

Williams, Raymond, *The Country and the City* (London: Chatto and Windus, 1973).

Wood, Mrs Henry (Ellen Price), *The Surgeon's Daughters* in *Lady Grace and Other Stories*, 3 vols (London: Richard Bentley & Son 1887), III, pp. 45–167.

Young-Zook, Monica M., 'Wilkie Collins's Gwilt-y Conscience: Gender and Colonialism in *Armadale*', *Victorian Sensations: Essays on a Scandalous Genre*, eds. Kimberly Harrison and Richard Fontina (Colombus: Ohio State University Press, 2006), pp. 234–45.

CHAPTER 5

Physiognomy, Evolution, and the Divine in Ruritanian Romances

Introduction

In his 1892 work, *Degeneration*, Max Nordau scoffed at the late nineteenth-century attitude he defined as the 'vague qualms of a Dusk of the Nations, in which all suns and all stars are gradually waning, and mankind with all its institutions and creations is perishing in the midst of a dying world'.[1] These concerns about the impact of time, history, and the impending future on a Victorian present have largely, though sometimes too strongly, become synonymous with modern perceptions of the Victorian *fin de siècle*: 'a general anxiety about the nature of human identity [....] an anxiety generated by scientific discourses, biological and sociomedical, which served to dismantle conventional notions of "the human."'[2] While I hesitate to generalise that all Victorians suffered from *fin de siècle* fears, and certainly not to the exclusion of other decades in which the same fears were evidenced (as seen in particular in Chap. 4 examination of mid-century proto-eugenics and obsession with lineage), one cannot deny the presence of *fin de siècle* anxiety, which was in part sparked from or exacerbated by a deluge of texts concerning the body, human genesis, and social divisions. The texts which particularly aggravated these concerns—and which will be the focus and the theoretical backbone of this chapter—are the works of Charles Darwin.[3] An interrogation of the works of Darwin on late nineteenth-century literature is

A. Boucher, *Science, Medicine, and Aristocratic Lineage in Victorian Popular Fiction*, Palgrave Studies in Literature, Science and Medicine, https://doi.org/10.1007/978-3-031-41141-0_5

hardly ground-breaking; it is, however, an excellent place to start when it comes to the two wildly understudied genres that serve as the basis for my next two linked sister-chapters, which I will discuss below.

Part of the response to *fin de siècle* fears was a reinvigoration of the Medieval Revival which, as Victorian Medievalism critics Holloway and Palmgren classify it, was 'an anchor in a time of stormy upheaval'.[4] Medievalism as a literary and aesthetic movement had existed since the eighteenth century, albeit under many different guises, and ghosted in and out of vogue for much of the Victorian era. It found further and vigorous life working with and against the inundation of *fin de siècle* tensions.[5] A 'one size fits all medievalism', the Medieval Revival allowed artists and audiences 'to go beyond or to completely dismiss true historical study of the period to focus on what fit his/her current imagination and taste'.[6] In particular, the Medieval Revival became another method for those at the end of the nineteenth century to contemplate human origin. It provided comfort and a way to return to a glorified past; it was used to exploit fears of degeneration; it was the locus for social commentary, criticism and philosophy; and it at times espoused a clinical and scientific view about how society had arrived at the contemporary moment from a historical starting-point.

Returning to concepts or constructions of the medieval invariably means investigating concepts or constructions of feudalism; arguments surrounding the aristocracy and representations of aristocratic lineage became inextricable from many *fin de siècle* Medievalist approaches to nostalgia, (d)evolution, civilisation, human nascency, and the future.[7] Historian Norbert Elias posits in his work, *The Civilizing Process*, that modern treatments of the medieval aristocracy tend to be little more than conduits of contemporary feeling:

> Whether the medieval warrior came to be seen as the "noble knight" (only the grand, beautiful, adventurous and moving aspects of his life being remembered) or as the "feudal lord", the oppressor of peasants (only the savage, cruel, barbaric aspects of his life being emphasized), the simple picture of the actual life of this class is usually distorted by values and nostalgia from the period of the observer.[8]

This chapter examines Ruritanian romances, a genre of fiction very concerned with Elias's 'noble knight', an aristocrat possessing an unscientific, hereditarily 'pure', and unevolving body. The chapter that follows looks at a genre, Evolutionary Feudal fiction, which still felt the need for an interrogation of the medieval but went in a contradictory direction from Ruritania, to look instead at Elias's 'feudal lord', whose body is rooted in a changing, scientific, and animal world. Elias's emphasis on a mutual distortion and nostalgia is a thread which has run through all chapters and genres in this monograph but is especially resonant here, where—according to Elias—portrayals of the medieval world tend to be one of two binary options. As discussed in the introduction, these are the only two genres or chapters which have a complete overlap in timescale, and theoretical framework. Despite, or perhaps because of, these genres supposedly polarising representations of aristocratic evolution and their use of the medieval period, the genres are actually bound more tightly together than their initial premises may allow—reactionary, in dialogue, and deeply affected by the scientific and medical discourse which had been accumulating in literature and culture for the last century. In both of these genres do we see a sort of breaking point reached, with one genre fleeing into denial and the other wallowing in despair. The former—Ruritanian romances—argues that aristocratic physicality and lineage is designed by a higher power for altruistic purposes; the latter—the Evolutionary Feudal—portrays aristocratic heredity as uncaringly carved by nature and elected to power through the vagaries of 'survival of the fittest'.

Both genres, but especially Ruritanian fiction, uncomfortably situate ideas of Darwinism next to theories espoused by the enormously popular pre-, early- and mid-Victorian philosopher Thomas Carlyle. Ruritanian romances particularly echo Carlyle's sentiments—in his 1841 work *On Heroes, Hero-Worship and the Heroic in History*, Carlyle rhapsodises over the belief that a true leader of men is divinely or cosmically appointed.[9] Not only did one of Carlyle's contemporaneous reviewers interpreted Carlyle's dogma as the belief that 'the progression of humanity depend[s] up on the veneration of the Divine in man' rather than on evolution or other forms of social and intellectual advancement, but Carlyle himself created a 'parody of the visions of science from the early reform era' in his *Sartor Resartus*, taking a more spiritual approach to the study of the natural world and the history of the elite.[10] In fact, Charles Darwin himself stated 'that he had never met anyone less suited to scientific enquiry than Carlyle', although Carlyle would likely say the same about Darwin and metaphysical enquiry.[11] As dozens, if not hundreds, of critics (both contemporary and contemporaneous) have noted, Darwin's theories do not 'eradicate deity altogether [but] assuredly removes it from any daily concern with the affairs of humanity'.[12]

One of Carlyle's primary philosophical topics was the nature of leadership and the desirable qualities of an aristocracy, which would, of course, become much more greatly complicated twenty-to-thirty years later when mingled with the more practical concerns and mechanics of evolutionary biology, socio-economics, and political reform. Carlyle's critique of modern government for its lack of true command first appeared in his 1831 text *Sartor Resartus*, where Carlyle argues that 'a King rules by divine right. He carries in him an authority from God, or man will never give it him [... H]e who is to be my Ruler, whose will is to be higher than my will, was chosen for me in Heaven.'[13] This sentiment was supplemented by his 1837 *French Revolution*, was fully expanded as the thesis of his 1841 lecture-series-turned-book, *On Heroes and Hero-Worship*, and continued to be expressed in his 1850 *Latter-Day Pamphlets* and his 1858–1865 *Frederick the Great*, before finding itself as the unlikely philosophical base of breezy *fin de siècle* adventure stories or the metaphysical antithesis to a particularly dispiriting branch of speculative fiction.[14] And yet perhaps this connection was unavoidable, no matter the decades between Carlyle and the new genres: in *On Heroes and Hero-Worship* he scrutinises the development of the aristocracy from tribal man to his present day, while still divorcing the true leader from time through his or her perpetual election and re-election by a higher power. *How* a true leader is appointed is always the same—a cosmic appointment—but the shape of that leader is decided by the needs of his or her era. Chris R. Vanden Bossche goes so far as to argue that Carlyle 'attempts to escape history' entirely which, as will be examined, makes Carlylean discourse a perfect thematic foundation for Ruritania texts, while enriching some of the transhistorical debates seen in the Evolutionary Feudal.[15]

Meanwhile, the connection between Darwin's theories of evolution and the role of aristocracy not only is explicitly discussed by Darwin several times over the course of his arguments, but was also treated as implicit by some contemporary commentators. One such satirical piece in 1879, entitled 'The Origin of Man' and ostensibly written by 'Darwin', gives the narrative of the Duchess de Chimpanzee whose daughter, Lady Adeliza, has evolved so remarkably as to attract the marital attentions of the Prince d'Orang-Outang. 'We are advancing, my dear', states the duchess, 'You are whiter than I am. You can talk in your youth; I could not until middle age. Your grandmother, as you know, can only grunt it. You are moving to a higher sphere.'[16] Lady Adeliza is given Ethiopia as a dowry, marries the prince, and bears him a hairless, white son so unlike any creature seen before that it is given a new name: Man. Clear correlations are made between evolutionary advancement and social advancement, between sexual selection and the high society marriage plot, between the capabilities of the physical body and one's social status and wealth. More significantly for Evolutionary Feudal writers, these connections illustrated the innate need to classify and categorise, and yet at the same time how fluid, how unstable, and ultimately how futile such boundaries and classifications could be.

In her seminal work on Darwinian impact on Victorian literature and culture, Gillian Beer states that the reason for Darwin's complete absorption into nineteenth-century consciousness was the result of 'a work which included more than the maker of it at the time knew, despite all that he *did* know [...] With varying degrees of self awareness they [Victorian novelists] have tested the extent to which it [Darwinian theory] can provide a determining fiction by which to read the world.'[17] While Darwin was hardly the first to theorise about evolution and the Victorians were already familiar with other major naturalist theories (from his own grandfather Erasmus Darwin, J.B. Lamarck, Charles Lyell, Alfred Russel Wallace, and Robert Chambers), Darwin's texts had a far more profound and wide-ranging impact than any evolutionary theorists before him.[18] Darwinism transformed from a mere scientific theory to a ubiquitous and versatile perspective through which one could interpret every human experience, including, as will be made apparent, the role of the aristocracy, the mechanics of its socio-economic and biological lineage, and the interpretation of the physical form. Much like Carlyle, Darwin's works affected far more than their purported realm: the theories in his 1859 *On the Origin of Species* and his 1871 *The Descent of Man* (hereafter referred to as *Origin* and *Descent*, respectively) far transcended the boundaries of the natural sciences and became hugely embedded in much of the Victorian psyche, as is exemplified by the sheer proportion of literature which utilises, contradicts, or even brushes against Darwinism. In fact, it is difficult to find a major piece of mid-to-late Victorian literature that *hasn't* been academically appraised from a Darwinist perspective.

While Darwin's theories are widely understood, for the sake of clarity I will shorthand his theory of Natural Selection, which is the 'preservation of favourable variations and the rejection of injurious variations' in inheritable traits.[19] Natural Selection comprises two parts, the first being 'survival of the fittest'. This phrase originated in Herbert Spencer's 1864 *The Principles of Biology*, and was then later borrowed by Darwin in subsequent editions of *Origin*.[20] 'Survival of the fittest', as Darwin employs it and as it will be understood in this book, supposes that organisms with the traits best suited to their environments will survive where others would die out.[21] The second part of Natural Selection is 'sexual selection', in which organisms which reproduce sexually will choose the mate with the traits best suited to their environment, thereby producing offspring which will stand the highest chance of survival. In this way, natural law determines which species thrive or go extinct, or ascend and descend a hierarchy within nature. Both *Origin* and *Descent* appeared at the tail-end of Carlyle's career and provided a different explanation for questions about time and history, elitism or 'natural' aristocracy, and ways in which one could speculate about the institution's future, as both of these literary genres do at their core.

The structure of these two chapters will be slightly different than before. While I still attempt a slight survey of each genre, both of these genres are (at least in this period) considerably smaller than others I have analysed. Texts in these genres at the *fin de siècle* number perhaps one or two dozen, rather than in the hundreds. Even the Evolutionary Feudal, which would gain more traction in the mid-twentieth century with the rising popularity of apocalyptic fiction, would lose something of its essence and become far less concerned with tracing the roots of the aristocracy (although Darwinian competition and questions of leadership would always be a defining element). Put simply, there is less of a field for me to survey, so I have opted for the depth of a few case studies instead of breadth. Even more intriguingly (and appropriately), the major texts from each genre very noticeably build from their predecessors; by looking at a few texts individually and largely sequentially, one can watch a genre *evolve*, with its formal structures devices interacting with or mirroring the very evolutionary logistics it critiques.

Ruritania, or the Chivalric Feudal

Anthony Hope's 1894 novel, *The Prisoner of Zenda*, is a high-romance adventure story which stemmed from both a cluster of *fin de siècle* concerns and a chivalric resurgence. Set in the fictional country of Ruritania, after which the genre was named, *Zenda* sparked a handful of sequels and imitators, although few could capture the popularity of the original.[22] The typical plot of Ruritanian fiction focuses on a wealthy or aristocratic individual from England or America who stumbles into an unfamiliar Germanic or Balkan principality. He or she is surrounded by monarchs and nobles who react to the rotating set pieces of mistaken identity, court intrigue, and a threat to their nation's throne. The protagonist embarks on an ideological rebirth, shrugging off the cynicism, ennui, and bad behaviour of modern times as he or she becomes more embedded in the dignity of the Ruritanian world, creating a heavily derivative and decidedly pro-aristocratic space in late-Victorian and early-Edwardian literature. Despite *Zenda*'s popularity (as evidence by the book's thirty reprints in the two years after its original publication), all the texts in the Ruritanian genre, including *Zenda*, have been almost completely ignored by critics from the original publication to the present.[23]

The genre celebrates the romantic uncomplications of a place and history that never existed, and a highly simplistic aristocratic heredity—a sort of uchronian fiction rather than utopian. The few critics who have dealt with the subject of Ruritania at any length tend to analyse the genre in terms of empire and orientalism.[24] I argue something else entirely: that Ruritania, at least in its origin in *Zenda*, fetishises history rather than location. Despite Ruritania being a globally insignificant, non-English-speaking, Catholic absolute monarchy located hundreds of miles from Great Britain, it is easily read as a stand-in for the fantasy of merry old England. Nyman argues that by 'locating its action on the borders of the West and imaging an Eastern European kingdom named Ruritania it […] follows the conventions of the truly orientalising adventure narratives by such writers as H. Rider Haggard and G.A. Henty'.[25] While Orientalism may be an element of other Ruritanian fiction, Hope's Ruritania is not set on the borders of the West, but rather firmly inside it.

Though most Ruritanian texts had a contemporary setting (the earliest being set only 1730, and never in the actual medieval period), there is a purposeful vagueness in dating the writing. This ambiguity of tone allowed authors to depict absolute monarchies in glamorised, revisionist ways by combining nostalgia for pre-industrial society with the mores and morality of the authors' time and generally middle-class perspective. In doing so, Ruritania avoids Norbert Elias's contention that '[i]f members of present-day Western civilized society were to find themselves suddenly transported into a past epoch of their own society, such as the medieval-feudal period, they would find there much that they esteem "uncivilized" in other societies today'.[26] In Ruritanian romances, the ideal monarch is honest, hard-working, respectful, gender-normative, and is easily identifiable as such through bodily physiognomy which is highly visible to his or her subjects.

They operate in the same modes of masculine health and morality as G.W.M. Reynolds's diligent and virtuous working-class men—as if these manly exemplars from Chartist fiction could somehow be born to their titles (instead of being awarded a title for avoiding the very debaucheries of that system which will now one day corrupt them). This Ruritanian physiognomy is passed down in perfect facsimile to the monarch's descendants, ignoring any reality of evolution or the operations of hard heredity. Even soft heredity scarcely seems to play a part, given that the slightest fluctuations of behaviour and morality (and they *are* slight) are often resolved quickly and with seemingly little effect on descendants.

As an extension to the positive aristocratic bodies of Ruritanian monarchs, peasants live healthy, merry, agrarian lives under a gentle feudalism which never oppresses them, but rather renders them loyal to their rightful rulers. Politics are non-existent, apart from the occasional plot by a stock villain to steal the throne or coerce a princess into marriage. In one of the more literalised depictions of the 'king's two bodies', dynasties must be preserved: the land and people suffer if the correct bloodline is not on the throne or if that bloodline is corrupted by decadent, weakening influences. Inga Bryden argues that Victorian Medievalism and, by extension, Ruritanian literature 'had two major aspects: naturalism, which equated the past with simpler modes of feeling and heroic codes of action, and feudalism, which regarded earlier social structures as harmonious and stable'.[27] These views, as expressed in a Ruritanian setting, are not a Victorian call for reform but are rather a gently idealistic expression of mores typically seen in fairy tales; the purpose of Ruritanian fiction is not to foster harsh criticism (even towards Darwinian evolution), but to indulge in escapism and allow its readers respite or, as Nicholas Daly writes, to offer 'an escape hatch from modernity'.[28] When the protagonist of *The Prisoner of Zenda* first arrives in Ruritania, he chooses to nap in a glade instead of running for his train, in a conscious decision to appreciate pastoral history over the modern urban. The land itself rapidly becomes a dreamscape: 'To remember a train in such a spot would have been rank sacrilege. Instead of that, I fell to dreaming that I was married to the Princess Flavia and dwelt in the Castle of Zenda, and beguiled whole days with my love in the glades of the forest—which made a very pleasant dream', a dream that will, at least in part, come astonishingly to life during his visit.[29]

Ruritanian fiction attempts to pare down multifaceted middle-class Victorian views on the nobility: aristocrats are desirable leaders, but only if they conform to middle-class Victorian values, assure the public that those values will be continued in their descendants, and thereby deserve their position of social and biological supremacy. In Ruritania, one sees the echoes of aristocratic portraiture from genres past: the glamour and desirability of aristocrats in the silver fork novels, the middle-class moral, gender, and body normativity celebrated in Reynolds's work, and a security and knowability of the self (which was so conspicuously absent in sensation fiction). Ruritania posits that this formula, if executed properly, would leave lower-class citizens free to venerate the aristocracy without guilt or the desire for reform.

This thoroughly unachievable synthesis of desirable aristocratic characteristics is eventually taken further in Ruritania by the genre's suggestion that these aristocrats should strive to remain outside of the evolutionary scheme by producing children as physiognomically indistinguishable from themselves as possible, ensuring that a 'natural' aristocracy remains forever through untainted primogeniture. The genre imagines a utopia where classes exist but class tension does not, where heredity exists but evolution does not. If one could return to the feudal system and perfect it with anachronistic middle-class mores, as this hypothetical world has done, each class would successfully follow through on its duties and thereby erase internal politics and oppression.

In the reality of Ruritania, if one's moral and ancestral features manifest themselves physically then one's character and class could never be mistaken; physiognomy is an unquestionable fact. In this genre, aristocrats as a whole are able to be classified through external markings. When a leader is no longer capable or virtuous—when he or she is no longer truly *noble*, as Carlyle would consider the term—his or her body reflects it. Physiognomy had long been practised in the Western world and while Victorian medical philosophy often supported the idea that one's morality and class were physically conspicuous; Sharrona Pearl argues: 'Far more important than *being* of a certain class was *looking* like you were.'[30] Ruritanian fiction exaggerates this rhetoric as part of its uchronian construction away from realism and Darwinism—perhaps as allusion to the chivalric medieval works to which it is indebted. In part, this is due to the medieval tradition from which Ruritanian fiction is working: Graeme Tytler, scholar of physiognomy in the European novel, writes, 'Physiognomy in medieval literature is generally simple, and is confined mostly to references to family resemblances, nobility of features, pathognomic expressions and gestures, and, occasionally, the deceptiveness of the face.'[31]

The three works of Ruritanian fiction analysed below are Hope's *The Prisoner of Zenda* (1894), Robert Louis Stevenson's *Prince Otto* (1885), and Frances Hodgson Burnett's *The Lost Prince* (1915). These three texts approach the genre from different vantages, and yet arrive at the same conclusion which has come to stand as an overt theme in Ruritanian fiction: that subjects of the nobility want to love them, if only they and their offspring for generations to come can prove unwaveringly deserving. Or, in Carlylean terms,

> Society is founded on Hero-worship. All dignities of rank, on which human association rests, are what we may call a *Hero*archy (Government of Heroes)—or a Hierarchy, for it is 'sacred' enough withal! The Duke means *Dux*, Leader. King is *Kön-ning*, *Kan-ning*, Man that *knows* or *cans*, [that the] Hero is he who lives [...] in the True, Divine and Eternal, which exists always.[32]

Stevenson writes in *Prince Otto* that a leader should be 'a man of a courtly manner, possessed of the double art to ingratiate and to command; receptive, accommodating, seductive'; that, as Burnett argues in *The Lost Prince*, people 'are impressionable creatures, and they know a leader when they see him'; and, as Hope says in his prequel to *Zenda*, *The Heart of Princess Osra* (1896), that a good aristocrat represents to the people 'some sweet image under whose name they fondly group all the virtues and the charms'.[33] These quotations are in direct contradiction to Darwin's colder and more imprecise theories on the sliding scales of class and civilisation: 'Differences', Darwin writes, 'between the highest men of the highest races and the lowest savages, are connected by the finest gradations. Therefore it is possible that they might pass and be developed into each other'.[34] Despite the potential that evolution held for civilisation, society, and even the individual to continually progress and 'climb the ladder', there was also the potential for devolution; therefore, Ruritania's clearly imposed but benign class boundaries serve as a mode of middle-class wish-fulfilment for upper-class access without the hand-in-hand risk of returning downwards to a lower tier.

The Prisoner of Zenda

When exploring how Ruritanian novels portray the aristocratic body, physiognomic portraiture and divine physical inheritance become the two crucial tools for gauging an aristocrat's suitability to rule. The authors of Ruritanian works grant such authority to these tools that eventually any concept of realistic Darwinian heredity or influence of nature is completely abandoned by the text. An early reviewer of *The Prisoner of Zenda* gushes, 'That blessed word "Heredity" is likely to occur to the reader of the first few pages [of the novel]; but the thing itself, the pseudo-scientific thing, lifts not its horrid head and multiple issues for a single page to chill the romantic spirit.'[35] In his analysis of *The Prisoner of Zenda* and Hope's engagement with the 1893 heredity debates, Paul J. Niemeyer writes,

> In *The Prisoner of Zenda*, red hair and royalty are in fact inseparable parts of Rassendyll's inviolable biological inheritance—one physical, the other behavioral. By attributing the novel's success to readers' interest in biological descent, Hope suggests that the novel centers on the issue of how heredity determines the roles people play in life—specifically, those played by Englishmen in what was to Hope the modern world—and hints at the novel's possible origin in the scientific and popular discourse of 1890s London.[36]

I argue, however, that despite Hope's clear interaction with contemporary concepts of heredity, and despite physiognomy's ubiquity in Ruritanian texts, their function in this genre is more Carlylean than scientific: rather than making any scientific claims about heredity or physiognomy, Hope and other Ruritanian authors used them as familiar, easy methods of illustrating to the reader that a character is cosmically appointed to rule, or not. Physiognomy does little more in these texts than to bring about empirical proof inside the reality of the narrative of the presence of a divine hero-aristocrat. Hope does not challenge or address physiognomy directly, but rather obliquely. By operating inside a literary culture which made frequent use of physiognomic portraiture, Hope avoids dignifying scientific discussions of physiognomy, heredity, and evolution by giving them a response at all.

The Prisoner of Zenda's protagonist, Rudolf Rassendyll, presents one of the more visually striking manifestations of physiognomic heredity in the form of red hair and a long nose. Rudolf is the dissolute younger brother of a virtuous earl, both of whom are the descendants of a Ruritanian Crown Prince, Rudolf Elphberg, who had an illicit relationship with a Rassendyll countess on a diplomatic visit to England several generations before. That prince had been 'marked (may be marred, it is not for me to say) by a somewhat unusually long, sharp and straight nose, and a mass of dark-red hair—in fact, the nose and the hair which have stamped the Elphbergs time out of mind'.[37] Starting with Sir Walter Scott's work, nineteenth-century novelists increasingly made 'conscious use of national physiognomies, especially in [...] historical fiction, where references to peculiarly national faces underline the patriotic themes and also serve as symbolic expressions of the entire history of a particular race or people'.[38] In this case, the Elphberg look is not only a representation of a family and a country, but of a time gone by, and of the nostalgia which classifies both person and place. The hair and nose are visible signifiers, indicating even in another country who the Elphbergs (and since that liaison, the Rassendylls) are and what they stand for. Victorian readers were familiar with the literary interpretation of physical attributes. Based on the descriptions of the Elphberg looks one could reasonably assume, from both the serious and satirical works on physiognomy that informed Victorian readerships, that Rudolf's long nose and red hair signified his authority and vitality, even if other characters in *Zenda* merely associate his appearance with an aristocratic affair from an earlier age.[39]

Rudolf's brother Robert, the earl, has escaped this taint by being born with dark hair and shorter nose, and he and his countess express distaste at the tenacity of Elphberg heredity: "'She [the countess] objects to my doing nothing and having red hair," said I [Rudolf] [...] "It generally crops out once in a generation," said my brother [...] "I wish they didn't crop out'".[40] That Rudolf automatically pairs his laziness with his hair colour shows a clear perception of appearances' entanglement with behaviour—a perception which Hope's text supports, but which is wrongly interpreted by the earl and countess in this specific instance. Robert has no contaminating traits particular to any mentioned family history, and his actions are therefore free from predestination. Robert is allowed to be a modern nobleman who 'rises at seven and works before breakfast', exhibiting the Protestant work ethic which the middle classes so valued, especially in men.[41] The standard that Robert sets for all English aristocrats champions evolution and adaptation, while being the antithesis of his red-headed brother Rudolf's *sprezzatura*, or natural grace of God. The implication, based on Robert's attitude towards the family's tenacious hereditary traits, is that he and his ancestors likely practised strict sexual selection to distance themselves and their offspring as far from the Elphberg genetics as possible. Yet little dilution has taken place. A greater power overrides nature and evolution, allowing a perfect Elphberg copy to appear 'once in a generation' and to keep that generation from ever fully moving forward from the past.[42] This is not to say that Hope is hostile towards evolutionary theory: unglamorous though he may be, Robert is a man who is portrayed as more suitable to Britain's socio-political climate at the *fin de siècle*, while Rudolf is imprisoned by the specific values of the aristocrat in the past whose actions became Rudolf's moral and physical genesis.

While Robert is as dedicated, abstemious, and virtuous a man as one could hope to find in power, his personality is unremarkable to the point of tedium. He is easy to like, but impossible to worship and, according to both Hope and Carlyle's standards, he is therefore not a real leader. Rudolf has all the potential of being a worthless, parasitic aristocrat, and yet Hope casts him as the protagonist and marks him out through physiognomic composition as a true nobleman. Sophie Gilmartin argues that these storytelling traditions predominantly feature aristocrats as protagonists and 'the reader expects that at birth the hero or heroine will be set apart in some way as unique or superior to others'.[43] Being a chivalric romance, Ruritanian fiction repeatedly brushes against the pseudo-medieval fairy tale tradition and, true to form, Rudolf possesses a superior ability in most things, especially in leadership, just as his physical appearance had predestined.

In order to escape his cynical, useless existence, he leaves England and addresses his origin in Ruritania which, like him, seems to be misplaced in time. He senses, as Virginia Zimmerman says in her work on the Victorian understanding of the past, *Excavating the Victorians* (2008), that the mining of history 'revealed the extraordinary power of certain items to endure' even as humanity proved its mutability over time.[44] In sensing this, Rudolf understands that his identity, which is rooted in history, may still be available to him in the present. Once in Ruritania, Rudolf runs into his distant relation, the new King of Ruritania, and both men discover that the Elphberg genetics have cropped up strongly in them, rendering them almost twins in looks and conduct. Hope reinforces this point by naming them both (as well as their common ancestor) 'Rudolf'. This high level of genetic inheritance takes Darwin's supposition that '[h]abits, moreover, followed during many generations probably tend to be inherited' to an absurd level, to the point where nature is no longer in control, but rather a higher power is.[45] Their looks and personalities contain no sense of proportion of their genetic proximity, nor take into account any learned behaviour, nor the general probability of inheriting identical traits. Hope ignores all these scientific provisos in favour of an intrinsic, undiluted and unbreakable physical mirroring and code of conduct between two very distant cousins who have never met. Niemeyer reads these genetic similarities as Hope's 'comfort' with the ideas of German hard hereditarian August Weismann, discussed in the last chapter, 'who saw heredity as an ancestral presence manifesting itself in modern-day individuals', although the similarities are so improbable that they may very well be Hope's reaction against evolutionary concepts altogether.[46] Weismann *did* theorise that some traits could endure and crop up again over several generations, but it would be unlikely that *all* traits would do so, as is the case with the many Rudolfs in *Zenda*.

When King Rudolf is kidnapped by his younger brother in an attempt to steal the throne, Rudolf Rassendyll secretly takes the king's place while a rescue mission is planned, in order to keep the public from panicking. When Rudolf is finally put to good use in the correct, historically tinged environment, his decadent actions are transformed into chivalric, noble ones. He rescues the real king, foils the villain and wins the love of the people as he completes his Carlylean transformation in which the embryonic hero reveals his true nature: 'he must march forward, and quit himself like a man,—trusting imperturbably in the appointment and *choice* of the upper Powers [i.e. God].'[47]

As Rudolf becomes more embedded in his duty to his true, spiritual homeland, his hair and nose cease to be referenced. In *Zenda*'s sequel, *Rupert of Hentzau*, Rudolf's appearance is scarcely mentioned at all, and then only as a means to carry out another 'mistaken identity' plot. While the traits themselves do not fade, their significance does. In England, his Elphberg appearance marked him out as the descendent of an affair, with all of the moral trappings which go along with such a reminder. In Ruritania, his Elphberg appearance marks him out purely as an Elphberg, and as such, he is thought to be a legitimate ruler. Since his bloodline is linked to the land and his traits are now in their correct environment and symbolic time period, he is finally allowed to embody what those traits really mean.

Like the readers themselves, Rudolf is forced to leave Ruritania at the end of *The Prisoner of Zenda* and return to Britain. That he cannot stay in the land which his identity and physicality have gone so deeply into reifying is a commentary on the pleasant but ultimately impractical nature of viewing the medieval period nostalgically, although all other dynamics in the novel do everything to reify a nostalgic reading—including the eventual presence of both a prequel *and* a sequel. In the novel's sequel, *Rupert of Hentzau* the Ruritanian royal line ultimately fails, and the country is left to suffer the ravages of the modern world. Some of the last words in *Hentzau* read, 'Times change for all of us. The roaring flood of youth goes by, and the stream of life sinks into a quiet flow.'[48]

Prince Otto

Although Robert Louis Stevenson's 1885 proto-Ruritanian novel, *Prince Otto*, was published almost ten years before *Zenda*, it can be retroactively classified as part of the genre. To put it in more evolutionary terms, *Prince Otto* operates as a very slightly undeveloped or underformed *Zenda*. In fact, the unusual looks of the protagonists in both novels are so strikingly similar (and so wrapped up in discussing the ties between the moral and the physiognomic) that Prince Otto may be read in multiple ways as a slow-to-evolve ancestor to Rudolf Rassendyll.

While *Otto* is less a swashbuckling adventure and more a serious contemplation on leadership, public opinion, and the Medieval Revival, it set the groundwork for Ruritania and drew the initial parameters that *Zenda* redrew for subsequent fantasy romances. One of the most important elements explored by both Stevenson and Hope was the connection between aristocratic physiognomy and leadership. However, instead of treating the connection as a biological reality like Hope does, Stevenson looks at the connection as a matter of public opinion, not one of divinity, which is a more realistic perspective, but still divorced from Darwin's models of leadership and physical appearance. That Stevenson depicts the more subjective, interpretive side of physiognomy does not diminish its significance, but underscores the Carlylean need for a leader to exhibit correct behaviour to his or her public to earn the right to rule. Carlyle writes:

> We have undertaken to discourse here for a little on Great Men, their manner of appearance in our world's business, how they have shaped themselves in the world's history, what ideas men formed of them, what work they did;—on Heroes, namely, and on their reception and performance.[49]

While God is ultimately in control in crafting and placing an aristocratic body in a position of leadership, there is still a public element to that leadership, a 'reception and performance'.

The narrative follows the titular Otto, monarch of the feudal German principality of Grünewald.[50] Otto is a lazy and ineffective ruler, much disdained by his people, who spent most of his reign eschewing duty in favour of pleasure. Slightly echoing G.W.M. Reynolds's moral masculinity, Stevenson writes, 'He hunts, and he dresses very prettily—which is a thing to be ashamed of.'[51] Otto foists all responsibilities on his wife, Princess Seraphina, and his councillor, Gondremark, who is 'the hope of Grünewald [...] he's a downright modern man—a man of the new lights and the progress of the age'.[52] Much like Rudolf's brother in *Zenda*, the highly evolved Robert, Gondremark is the champion of progress and the enemy of the nostalgic past. This antagonism of the future is far more clearly delineated in *Prince Otto*, since Gondremark is the undisputed foe of the charming protagonist, as it is rumoured that Seraphina and Gondremark are lovers and that he has designs on the throne. Gondremark is the embodiment of the *fin de siècle*: an unstoppable charge into the future, with equal potential for utopic or catastrophic results. This tension between the old and the new creates an identity vacuum for the people, who are undecided whether they wish to remain in a Ruritanian landscape at all; in terms of the body politic, the king seemingly *has* no body.

A large part of the populace's wish to leave that Medievalist landscape is due to their uncertainty about their ruler's physicality. Since Otto lives a private life instead of being a public figurehead, no one can tell whether his body is altruistic and cosmically appointed to the role, or if it is selfish and degenerative. Otto's subjects are rife with speculative codings of the aristocratic body and indicate their urgent need to 'read' and interpret Otto; however, Otto refuses to present himself as a text and his absence forces his subjects to imagine the worst. They theorise about his physical form, arguing that Otto surely must be bald and sickly looking, since those physiognomic traits would explain his poor leadership.[53] The reader, who not only can see Otto, but who is also indoctrinated in the long literary tradition of physiognomic portraiture, interprets Otto's actual form (which is tall, handsome, healthy, and with a head full of red curls) not as a critique of the inaccurate nature of physiognomy, but rather as a romanticised reinforcement of it: the reader knows that Otto is full of good intentions and untapped potential, based not only on his role as a sympathetic protagonist but also because he is handsome.

In one instance, this exaggerated physiognomic portraiture goes so far as to challenge ideas of Darwinian degeneration. Otto is described, this time by a courtier who has seen him in person, as a man with

> hair of a ruddy gold, which naturally curls, and his eyes are dark, a combination which I always regard as the mark of some congenital deficiency, physi-

> cal or moral [...] His one manly taste is for the chase. In sum, he is but a plexus of weaknesses; the singing chambermaid of the stage, tricked out in man's apparel, and mounted on a circus horse.[54]

The audience knows from its personal experience with Otto that he is not weak or a 'singing chambermaid', and this knowledge discredits the initial claims that Otto is suffering from a deficiency or degeneration. The very mention of such speculations underscores the unpleasantness of *fin de siècle* degenerative fears, how out-of-place they appear in this novel and its Medievalist setting, and how little those fears deserve to be applied to the protagonist. This is but one manifestation of how, as Menikoff writes, Stevenson 'explored the matter of faith in an age of evolutionary biology', in this instance pushing faith in Otto (and by extension, God, for placing Otto on the throne) onto the reader.[55] Stevenson would, of course, go on to write his novella *The Strange Case of Dr Jekyll and Mr Hyde* the following year, which gives voice to explicit fears of degeneration.[56] However, *Prince Otto*, which he subtitled 'A Romance', may have been the initial way in which Stevenson addressed *fin de siècle* uneasiness about degeneration, namely by attempting to ignore it in favour of a romanticised model for physical appearance, rather than the degenerative or Darwinian model. Stevenson himself acknowledged (as quoted by a reviewer) that the novel has '"a wonton [*sic*] air of unreality"; and he [Stevenson] puts it down to "the difficulty of being ideal in an age of realism"'.[57] The unrealistic model represented in *Prince Otto* is one in which morality perfectly aligns with outward appearance, the only real (d)evolution being in public opinion. The aristocracy may not evolve in this novel, but the public still needs to see a leader in order to be reassured that their faith in said leader is well-placed.

The people of Grünewaldian are desperate for a hero upon whom they can construct their national identity. When Otto refuses to become this figure by staying out of the public gaze, they turn instead to a Republican movement which favours Gondremark as their natural leader and allows for its members to be easily identified through the medals they wear: 'drawing out a green ribbon that he wore about his neck, he held up [...] a pewter medal bearing the imprint of a Phoenix and the legend *Libertas*.'[58] What is important to this movement is not necessarily the ideology, since there are only two perceived options in the political environment (Otto and thus far inadequate traditionalism versus Gondremark and probably beneficial change), but rather the method by which the Republicans show their solidarity: through what they wear. By wearing a mass-produced medal, one carries around one's hopes, beliefs and prejudices on one's body, to be easily understood by anyone who recognises it, which, in the case of Grünewald, is a high percentage of the population. The medals are depicted as an insincere and desperate expression of identity and creed arising from the lack of any alternative, a false physiognomy to fill the void that a true leader's physiognomy would occupy. The medals symbolically reinforce the people's need not only for a country-wide physiognomic understanding in the wake of Otto's physical absence, but also for their need for glory and for a divine leader. Of all the potential external markers to wear to signal political allegiance, a medal—typically earned through merit, service, or heroism—is an intriguing choice. The wearer of an *earned* medal is usually entitled to a degree of accolade and, in the public's turn to Republicanism, every citizen becomes his or her own hero with his or her own medal, given freely instead of merited. Instead of building their own national identities around a God-given Carlylean leader or national champion who provides some degree of glory, envy, and aspiration to the masses, every citizen may worship his or herself. While Stevenson by no means criticises Republicanism, his narrative does support Carlyle's assertion about leaders as standards for others:

> He [a king] is practically the summary for us of *all* the various figures of Heroism; Priest, Teacher, whatsoever of earthly or of spiritual dignity we can fancy to reside in a man, embodies itself here, to *command* over us, to furnish us with constant practical teaching, to tell us for the day and hour what we are to *do*. He is called *Rex*, Regneator, *Roi*: our own name is still better; King, *Könning*, which means *Can*-ning, Able-man.[59]

What Stevenson attempts to do through his involvement through his engagement with socio-scientific discourse on leadership is to destigmatise the desire to worship elite members of society. Stevenson argues through his text that if middle-class virtue can be exhibited by the nobility and their worth can be proven to the masses, the public will be able to satisfy a very real need without guilt; that what the middle class objects to is not an aristocracy itself, but to weakness and worthlessness in rulers. The lower classes of Grünewald are 'proud of their hard hands, proud of their shrewd ignorance and almost savage lore, [and] looked with an unfeigned contempt on the soft character and manners of the sovereign race'.[60] Otto is unable to provide the necessary model for his people. Public opinion eventually wins and Grünewald is thrown into revolution. Otto and Seraphina reunite but are forced to escape into exile to live peaceful, private lives at a foreign court and presumably never have children, whose soft heredity an audience would have been able to assess in terms of Otto's moral rejuvenation.

Grünewald is in many ways the anti-Ruritania. It is not a place lost to time, but a place that is catching up with time, evolving into a modern nation—although modernity is correlated in this context with violence, misunderstanding, and a distinct *fin de siècle* sense of degeneration finally coming to fruition. Modernity is the antagonist in the novel and, in some capacity, the antagonist in all Ruritanian fiction while the doomed feudal days are presented with bucolic, nostalgic poetry. Stevenson punishes the country of Grünewald for modernising and for not respecting or reacting to Otto's personal growth: the country forcibly ejects a leader who the reader knows to be good, having now fully aligned with modern values that are pleasing to the audience; the violent upheaval is unnecessary, revealed to be ultimately unsuccessful, and leads to the total destruction of the nation. Stevenson introduces the book by saying:

> You shall seek in vain upon the map of Europe for the bygone state of Grünewald. An independent principality, an infinitesimal member of the German Empire, she played, for several centuries, her part in the discord of Europe; and, at last, in the *ripeness of time* and at the spiriting of several bald diplomatists, vanished like a morning ghost.[61]

Without its leader, which Carlyle states is an 'eternal corner stone, from which they [the people] can begin to build themselves up again', Grünewald ceases to exist, not just in essence or in name, but physically.[62] It is erased from the map, swallowed up by other lands, much like Otto who resides in another nation and has taken ownership of his private and therefore insignificant body and identity. There is no face for Grünewald, and if there is no face, there can be no physiognomy, further tying person to place, and physical nature to physical landscape. Much like Hope's slow extinction of Ruritania at the end of *Rupert of Hentzau*, Otto and Seraphina slowly but happily fade away in another mystical land, leaving their country to face its burgeoning modernity without the guidance of its aristocrats.

The Lost Prince

The Ruritanian text which most clearly correlates the body of the leader with the health of the land is Frances Hodgson Burnett's 1915 children's novel, *The Lost Prince*. This is also the text which is most heavily imbued with the notion of a natural aristocracy, or a human 'emblem of the Godlike, of some God'.[63] While there is a thirty-year gap between *The Lost Prince* and *Prince Otto* and a twenty-year gap between it and *The Prisoner of Zenda*, as well as a difference in intended audiences, Burnett's representations of the aristocratic body and heredity are still deeply embedded in discourse from decades previously. These long-standing tropes demonstrate, rather aptly, that there was little evolution in the Ruritanian genre.

The story follows Marco Loristan and his father, displaced citizens from the fictional Balkan country of Samavia, who travel the world quietly and keep the secret of the lost Samavian prince. Generations before, a mad Samavian king quarrelled with and stabbed his heir, Prince Ivor, who escaped both his violent father and any sense of hereditary taint (a plot which would have made the novel a suitable candidate for sensation fiction). The king was deposed and the country withered under the factions who warred repeatedly for the crown: 'From that time, the once splendid little kingdom was like a bone fought for by dogs. Its pastoral peace was forgotten [...] It assassinated kings and created new ones. No man was sure in his youth what ruler his maturity would live under.'[64] The Loristan family's duty is to track and protect the descendants of Prince Ivor and, when the time is right, raise the necessary support to place the hidden, legitimate king on the throne. Much as *Zenda* and *Otto* play with the notions of fairy tale, and through them idealise and glamorise the aristocratic body, so does *The Lost Prince*, which is the closest of all Ruritanian fiction to expressing ancient or primitive class superstitions as a modern reality or desire. James Frazer, in his work on the primitive superstitions of the upper classes, writes, 'His [a king's] person is considered [...] as the dynamical centre of the universe, from which lines of force radiate to all quarters of the heaven [*sic*]; so that any motion of his—the turning of his head, the lifting of his hand—instantaneously affects and may seriously disturb some part of nature.'[65] While Samavia is less of a stand-in for Great Britain than Ruritania and Grünewald had been (Samavia being a Balkan state instead of a German one), the legend of Ivor reinforces the connection to Great Britain through its parallels to the dethroned James II and his exiled heirs.[66] While Burnett is more focused on the optimism surrounding aristocracy rather than with commenting on Great Britain's monarchical history, the connection serves to make Samavia another familiar land which can host the nostalgia for a time gone by through legend and allusions to real historical events.

Just as *Zenda* had philosophised that certain blood produces certain traits, especially in terms of class-based or national identities, so *The Lost Prince* argues on a greater scale: the land itself manifests certain traits based on the blood of its ruler, and an altruistic body divorced from Darwinism is needed to rule that land effectively. Despite five hundred years of national strife, Marco is taught the simplistic lesson that Samavia is guaranteed to heal only when its true leader returns. The rigid adherence to a system of binaries (good/bad, rightful king/usurper, healthy land/dying land) supports the idea that blood and biology are tied to a place and class, underscoring Ruritanian fiction's repeated discussion regarding aristocratic physicality and the escapism it provides for its readers. In this instance, the complex and grim politics that emerged from the Baltic states at the time of Burnett's writing are ignored for simple and idealised ones, to accompany the simple and idealised aristocratic body. Despite the story of the mad king, the text maintains a firm belief in the rightness of rule through heredity, provided that heirs are bred and trained to be exact duplicates of their virtuous parent, again skewing Darwin's theories of heredity and inheritance to such preposterous degrees that the novel's examples of heredity no longer have any place inside of Darwin's arguments, nor in the realm of nature at all.

Burnett hints early on that Marco and his father are really the descendants of Prince Ivor, though the text does not admit it until the end of the novel. Marco might not realise his own status, but the audience (in its role as spectator within Ruritania's physiognomic, anti-Darwinian tradition) knows what Marco does not. What enlightens the reader is the constant emphasis on the ways in which Marco's blood and body perform: 'When they talked together of its [Samavia's] history, Marco's boy-blood burned and leaped in veins, and he always knew, by the look in his father's eyes, that his blood burned also.'[67] Even the idea of the country sparks a surge of vitality in him, as though the land and the person were two chemicals reacting to each other on a cosmic platform, far closer to the realm of Carlyle than to the realm of Darwin.

Samavia is manifested in the Loristans' physicality both internally and externally. Marco proves to be a natural leader, and is easily identified as such through the eyes of others. The Loristans' servant says of him, 'the young Master must carry himself less finely. It would be well to shuffle a little and slouch as if he were of the common people', though as far as Marco and presumably the reader are aware, he *is* of the 'common people' and has not been raised with any connection to high society.[68] The text assumes that aristocrats are of a different constitution, almost a different species, and it would be impossible for the naked eye to mistake one of their class, that 'all sorts of Heroes are intrinsically of the same material'.[69]

The Loristan markings are not as distinctive as those in *Zenda*; while this is likely just a logistical or aesthetic choice for Burnett, it also plays into the physiognomic idea that dark hair and eyes 'frequently belong to the physically or morally strong', which aptly matches both the high-mindedness and physical hardiness of the Loristans.[70] What is distinctive about the Loristans, however, is an overwhelming aura of masculinity and maturity. A real aristocrat, in Burnett's work, is hyper-manly, in part due to his moral strength and in part his ability to shoulder great responsibility. Where aristocratic markings in *Zenda* were largely gender-irrelevant (being restricted mostly to an unusual hair colour), here they tie more closely to the masculine and moral ideals of knighthood. Marco 'was the kind of boy people look at a second time [...] he was a very big boy—tall for his years, and with a particularly strong frame. His shoulders were broad and his arms and legs were long and powerful. He was quite used to hearing people say, as they glanced at him, "What a fine, big lad!"'[71] Of all the children in the neighbourhood, Marco is the largest for his years, the most well-shaped, and the one whose form most implicitly produces awe in those who see him. He is a miniature version of his father, who 'was a big man with a handsome, dark face [who] looked, somehow, as if he had been born to command armies, and as if no one would think of disobeying him'.[72] A focus on the Loristans' strong body-type and their ability to command emphasises what Dominic Lieven declares to be 'the aristocracy's traditionally foremost occupation, namely war'.[73] In his expression of love towards his father, Marco exclaims, 'Father [...] I love you! I wish you were a general and I might die in battle for you.'[74] In addition, their strong soldierly bodies reinforce the medieval component of Ruritanian fiction and underline that they are out of time with the rest of the world and are easily identified as such through their knightly body-type.[75]

Little reference is made to Marco's absent mother, or to any woman in the Loristan history, making each man's genesis appear to be an asexual cloning process during which the bloodline is not diluted, never strays from direct primogeniture inheritance, nor contains any influence apart from the hyper-masculine. Nothing else in Ruritanian literature presents so direct an attack on Darwinism's adherence to sexual selection, or so fanciful an imagining of the capabilities of the aristocratic body. Marco's father repeatedly says with proud regard to his son's upbringing, 'Here grows a man for Samavia', underscoring the notion that his son was grown and cultivated, not born, and that what Samavia needs is the true masculinity of aristocratic influence.[76]

Despite the Loristan's physical connection to Samavian land, their noble countenances are not so much a product of Samavia itself but rather of a happier, feudal age in general. Much as in *Zenda*, there is an exoticism of time instead of geography. When recounting the legend of the lost Prince Ivor, Burnett writes of Samavia, 'In those past centuries, its people had been of such great stature, physical beauty, and strength, that they had been like a race of noble giants [...] The simple courtesy of the poorest peasant was as stately as the manner of a noble.'[77] While the correct bloodline was on the throne, nobility and hereditary health were a part of the land and the physiognomic makeup of all its people. The 'race of noble giants', in addition, calls back to numerous religious or folkloric texts the world over, which posit a race of giants as a sort of ur-population, often defiled or driven off through the years—which, to some extent undoes the genre's resistance to the concept of degeneration as a modern and urban phenomenon. All other humans in the novel have been made diminutive except the Loristan men, whose pure blood preserves them from the ravages of evolution and keeps their physical stature within historical, 'noble giant' dimensions. There is no evolution in this novel—only devolution and the small capacity to recover from said devolution. The Loristans have a genetic stasis from which everyone else has devolved and to which everyone aspires to return.

Loristan masculinity is the marker of an older aristocratic archetype, and their non-Englishness is not so much a criticism of England, but of modern, industrialised nations and the ugliness and weakness those nations foster in their people through (d)evolution, as will be seen in the upcoming section on H.G. Wells's *The Time Machine* (1895). Urban devolution is hardly an unusual critique of modern life, given that, as Bryden argues, 'many medievalist evocations of the past, from Scott onwards, implicitly contrast a glamorous lost world with drab modernity'.[78] Drab modernity is epitomised in Marco's friend, a young English boy nicknamed 'Rat'. He, in direct opposition to Marco's natural aristocracy, is both a lord's son and a devolved, infirm, animalistic weakling. He spends his days on the streets avoiding his savage drunkard father, and his class is indistinguishable from that of the members of his urchin gang. He constantly reminds characters and audience alike that his father is a 'gentleman [...] I am a gentleman's son', though initially the reader can see no characterisation to support that statement.[79] Rat says to Marco, 'I wish I was your size! Are you a gentleman's son? You look as if you were', and yet Rat himself does not fit the physiognomic model to which he subscribes.[80] He is not only a small, feeble boy, but also aggressive, close-minded, and portrayed as literally low, since he is disabled and must pull himself around on a small cart. His noble traits have been hidden or eroded by the foulness of modern London, making him unfit to rule anything but his group of waifs. His nickname rather heavy-handedly also stands in as a manifestation of his physiognomic self.[81] The rat, like an industrialised city, is ruthless, opportunistic and dirty. Rat and Marco embody Ina Zweiniger-Bargielowska's theory on the *fin de siècle* masculine body:

> A beautiful, healthy, and fit male body was identified with hegemonic masculinity whereas countertypes such as the stunted, narrow-chested urban labourer or the obese, flabby businessman signified degeneration. Cultural pessimism about modernity gave rise to growing fears of racial degeneration and biological based rhetoric permeated social policy discourse from the 1880s.[82]

Where in medieval Samavia the purity of the land elevates the peasant to the physiognomic level of the aristocrat, in modern Britain the corruption and ugliness of industrialisation degrades the aristocrat to the level of the beast. Again, the blood of the rulers is tied to the land, or at least undergoes a correlating change.

Rat, through his exposure to Marco, is reminded of the superior physicality that an aristocrat should possess, and quickly becomes absorbed in the high-mindedness of Samavia. He quickly re-evolves, becoming gentle, selfless and brave, and is named Marco's 'aide-de-camp'.[83] Though Rat does not revert to using his given name nor regain the use of his legs (both Hodgson Burnett's manifestations of his place in the hierarchy under Marco), he insists upon using crutches instead of his cart, picking himself off the floor and developing strength in his other limbs.

When the time for the restoration of the Loristans comes, the physical presence of Marco's father is needed in Samavia in order to command the support of the land and its people: 'If they [the Samavian public] could see the man with Ivor's blood in his veins, they'd feel he had come back to them—risen from the dead.'[84] The Loristan body is so fast-acting upon the land that the revolution happens off-screen almost overnight, indicating the further presence of supernatural or divine help; this is not the behaviour of landscape in nature, nor of societal infrastructure in a normal state of recovery. Though only a few weeks or even days previously '[w]ar and hunger and anguish had left the country stunned and broken', once Marco's father is crowned the country palpably begins to heal: 'food and supplies of all things needed began to cross the frontier; the aid of the nations was bestowed'.[85] Such is the fantasy of an aristocrat's power that bounty and joy follow in his wake, making him Carlyle's 'Great Man' who is 'by the nature of him a son of Order, not of Disorder', or Frazer's medieval kings who 'possess the same gift of healing by touch'.[86]

Conclusion

Although Ruritanian fiction largely fell out of favour by the 1910s, Nicholas Daly traces the ability of Ruritanian texts to endure for more than a century—sometimes in near perfect form—much like the aristocrats who popular their pages. Even including satirical rewritings like George MacDonald Fraser's *Royal Flash* (1970), the Fran Drescher film *The Beautician and the Beast* (1997) or K.J. Charles's *The Henchmen of Zenda* (2018), there is a strong connection between physiognomic reading and leadership ability. The most famous scene in the late Ruritanian film *The Princess Diaries* (2001) culminates in the 'big reveal' of a royal makeover, done at the insistence of the princess's glamorous grandmother before she dare debut the princess to the public. Unflattering, unsmiling pre-makeover pictures block the girl's face and are pulled back to reveal a glossy, beaming, and public-ready princess. That the heroine is played by a radiant Anne Hathaway—a challenging suspension of disbelief that the princess was 'ugly' and unsuitable—only underscores the film's perfect Ruritanian lineage. Much as the audience saw Prince Otto, Rudolf Rassendyll, and Marco Loristan's potential through their misunderstood physiognomy (which needed only the most minor of adjustments), so too does *The Princess Diaries* scarcely hide the Carlylean potential of the contemporary aristocrat—*the movie star*—in the looks of its protagonist.

Notes

1. Max Nordau, *Degeneration* (1892), 7th ed., translated from the 2nd ed. of the German work (New York: D. Appleton and Company, 1895), p. 2.
2. Kelly Hurley, *The Gothic Body* (Cambridge: Cambridge University Press, 1996), p. 5.
3. These texts include Charles Darwin's *On the Origin of Species* (1859), ed. by Jim Endersby (Cambridge: Cambridge University Press, 2009), and *The Descent of Man* (New York: D. Appleton and Company, 1871); Nordau's *Degeneration* (1892); Francis Galton's *Inquiries into Human Faculty* (1883) (New York; J.M. Dent, 1919), Edward Burnett Tylor's *Primitive Culture* (London: J. Murray, 1871), and James Frazer's *The Golden Bough* (1890), among just a few.
4. Lorretta M. Holloway and Jennifer A. Palmgren, 'Introduction', in *Beyond Arthurian Romances: The Reach of Victorian Medievalism*, ed. by Lorretta M. Holloway and Jennifer A. Palmgren (Houndmills, Basingstoke: Palgrave Macmillan, 2005), pp. 1–8 (p. 2).
5. Kevin L. Morris, *The Image of the Middle Ages in Romantic and Victorian Literature* (London: Croom Helm, 1984), p. 1.
6. Holloway and Palmgren, p. 2.
7. For an overview of the field of nineteenth-century Medievalism, see K. Morris's *The Image of the Middle Ages*; Elizabeth Fay's *Romantic Medievalism* (Houndmills, Basingstoke: Palgrave, 2002); Clare A. Simmons's *Reversing the Conquest: History and Myth in Nineteenth-Century British Literature* (New Brunswick, NJ: Rutgers University Press, 2006); and Holloway and Palmgren's *Beyond Arthurian Romances*, among just a few.
8. Norbert Elias, *The Civilizing Process*, trans. by Edmund Jephcott (1978), ed. by Eric Dunning, Johan Goudsblom and Stephen Mennell (Oxford: Blackwell Publishers, 1994, repr. 2000). p. 173.
9. For an overview of Carlyle's life and work, see Chris R. Vanden Bossche's *Carlyle and the Search for Authority* (Columbus, OH: Ohio State University Press, 1991); Simon Heffer's *Moral Desperado: A Life of Thomas Carlyle* (London: Weidenfeld and Nicolson, 1995); William Howie Wylie's *Thomas Carlyle: the man and his books* (London: T. Fisher Unwin, 1909), and D.J. Trela and Rodger L. Tarr's *The Critical Response to Thomas Carlyle's Major Works* (London: Greenwood Press, 1997).
10. Prof. August Tholuck, 'Review of *On Heroes and Hero-Worship*' (March 1847), in *The Critical Response to Thomas Carlyle' Major Works*, ed. by D.J. Trela and Rodger L. Tarr (Greenwood Press: London, 1997), pp. 100–02 (p. 101); James A. Secord, *Visions of Science* (Oxford: Oxford University Press, 2014), p. 234.
11. Secord, *Visions*, p. 234.

12. Secord, *Visions*, p. 234; Justin E.A. Busch, *The Utopian Vision of H.G. Wells* (Jefferson, North Carolina: McFarland & Company, Inc., 2009), p. 3.
13. Carlyle, *Sartor Resartus* (1834), eds. Kerry McSweeney and Peter Sabor (Oxford: Oxford University Press, 2008), p. 188.
14. The lecture series Carlyle gave for *On Heroes and Hero-Worship* was an immense success, selling out and drawing in famous attendees. Newspapers were eager to publish the lectures but Carlyle refused and made a larger profit publishing them in book form and a second edition was ordered the following year. Heffer, p. 203; p. 217.
15. Vanden Bossche, p. 98.
16. [Anon.] 'The Origin of Man', *The Bristol Mercury and Daily Post* (5 April 1879), 9637, p. 6.
17. Gillian Beer, *Darwin's Plots*, (London: Routledge & Kegan Paul, 1983), p. 4.
18. James A. Secord writes in *Visions of Science*, '"Darwinism", as it was quickly termed, was only the most prominent of the unifying projects that enhanced the explanatory power of science and its global ambitions' (p. 241).
19. Darwin, *Origin*, p. 71.
20. Herbert Spencer, *The Principles of Biology* (London: Williams and Norgate, 1864–67), II, p. 48; p. 53; p. 54; p. 58; p. 60; p. 90 etc.; Mark Francis, *Herbert Spencer and the Invention of Modern Life* (2007) (London: Routledge, 2014), p. 3; Mike Hawkins, *Social Darwinism in European and American Thought 1860–1946* (Cambridge: Cambridge University Press, 1997), p. 82; p. 84.
21. According to Mark Francis, one of the major differences between Spencer and Darwin in how the term 'survival of the fittest' was deployed was that Spencer believed an organism would be 'brought into general fitness with its environment and ... [u]nlike Darwin's theories, Spencer's ideas of growth and adaptation did not have *failure* as a primary focus. Instead they were part of a general belief in *success* measured by the number of adult life forms successfully living in harmony with their surroundings' (209).
22. Hope's *The Heart of Princess Osra* (London: Longmans, Green, and Co., 1896) and *Rupert of Hentzau* (1898) (London: Penguin, 2008), *Zenda*'s respective prequel and sequel, and his *Sophy of Kravonia* (1906), G.B. McCutcheon's *Graustark* series (1901–27), and Frances Hodgson Burnett's *The Lost Prince* (1915) (London: Heinemann, 1966), among a few.
23. Jopi Nyman, *Under English Eyes* (Amsterdam: Rodopi B.V., 2000), p. 42. A few exceptions may be found in Vesna Goldsworthy's *Inventing Ruritania* (New Haven, CT: Yale University Press, 1998); Raymond Wallace's 'Cardboard Kingdoms', *San Jose Studies*, 13:2 (1987), pp. 23–26; Paul J Niemeyer's 'The Royal Red-Headed Variant: *The Prisoner of Zenda* and the 1893 Heredity Debates', *College Literature* 41:1 (Winter 2015),

pp. 112–38; and Nicholas Daly's *Ruritania: A Cultural History, from* The Prisoner of Zenda *to* The Princess Diaries (Oxford: Oxford University Press, 2020).

24. Goldsworthy argues that Ruritanian authors 'exploited the resources of the Balkans [where later Ruritanian fiction was often set] to supply its [Great Britain's] literary and entertainment industries' (p. 2). Nyman supplements Goldsworthy's stance by defining Ruritania (using an early reviewer's words) as 'some semi-Oriental kingdom of Europe' (p. 41).
25. Nyman, p. 41.
26. Elias, *The Civilizing Process*, p. ix.
27. Inga Bryden, *Reinventing King Arthur* (Aldershot: Ashgate, 2005), pp. 33–34.
28. Daly, *Ruritania*, p. 4.
29. Hope, *Zenda*, pp. 21–22.
30. Sharrona Pearl, *About Faces* (Cambridge, MA: Harvard University Press, 2010), p. 37 [italics mine].
31. Graeme Tytler, *Physiognomy in the European Novel* (Princeton: Princeton University Press, 1982), p. 123.
32. Carlyle, *On Heroes and Hero-Worship and the Heroic in History* (1841) 2 vols (London, Macmillan and Co., Limited, 1926),, I, p. 14; II, p. 50.
33. Robert Louis Stevenson, *Prince Otto* (1885) (London: Chatto & Windus, 1912), p. 42; Burnett, p. 25; Hope, *Osra*, p. 335.
34. Darwin, *Descent*, p. 34.
35. Anonymous, 'Review of *The Prisoner of Zenda*', *Saturday Review of Politics, Literature, Science and Art,* 77:2013, 26 May 1894, pp. 556–57 (p. 556).
36. Niemeyer, p. 113.
37. Anthony Hope, *The Prisoner of Zenda* (1894), ed. by Tony Watkins (Oxford: Oxford University Press, 2009), p. 7.
38. Tytler, p. 235.
39. Eden Warwick [George Jabet], *Notes on Noses* (1848), 2nd ed. (London: Richard Bentley, 1864), p. 12; 'Physiognomy', *Anthropological Review*, 6:21 (April 1868), pp. 137–54 (pp. 143–44); Tytler argues, by way of contradiction, that nineteenth-century novelists preferred to use red hair to signify an evil character (p. 215).
40. Hope, *Zenda*, p. 6.
41. Ibid., p. 5; Danahay p. 7; The works of Thomas Carlyle and Samuel Smiles strongly espoused this ideology and were read and referenced very frequently in the middle classes.
42. Hope, *Zenda*, p. 6.
43. Sophie Gilmartin, *Ancestry and Narrative in Nineteenth-Century British Literature* (Cambridge: Cambridge University Press, 1998), p. 11.
44. Virginia Zimmerman, *Excavating Victorians* (Albany: State University of New York Press, 2008), p. 97.

45. Darwin, *Descent*, p. 157.
46. Niemeyer, p. 114.
47. Carlyle, *On Heroes and Hero-Worship*, I, p. 39.
48. Hope, *Hentzau*, p. 306.
49. Carlyle, *On Heroes and Hero-Worship*, I, p. 1.
50. Presumably meant to recall the area of Berlin named after Grünewald hunting lodge and forest.
51. Stevenson, *Otto*, p. 12.
52. Ibid., p. 17.
53. Ibid., p. 15.
54. Ibid., pp. 64–65.
55. Barry Menikoff, 'Prayers at Sunset' in *Robert Louis Stevenson Reconsidered*, ed. by William B. Jones, Jr (London: McFarland & Company, 2002), pp. 209–12 (p. 210).
56. Robert Louis Stevenson, *The Strange Case of Dr Jekyll and Mr Hyde and Other Tales* (1886), ed. by Roger Luckhurst (Oxford: Oxford University Press, 2006; reiss. 2008).
57. Q., 'Review of *Prince Otto*', *The Academy*, 1442, 23, December 1899, pp. 750 (p. 750).
58. Stevenson, *Otto*, p. 34.
59. Carlyle, *On Heroes and Hero-Worship*, II, p. 99.
60. Stevenson, *Otto*, p. 4.
61. Ibid., p. 3, emphasis mine.
62. Carlyle, *On Heroes and Hero-Worship*, I, p. 18.
63. Ibid., p. 12.
64. Burnett, p. 17.
65. Frazer, p. 202.
66. Perhaps the shift of setting from Germany to the Balkans was the result of Germany's modernisation and unification, making the forgotten, medieval wonderland of Ruritania no longer applicable to that part of the globe. Or maybe it was simply because the alliances and oppositions formed in WWI, at the time of Burnett's writing, would have made favouring a Germanic setting seem unpatriotic.
67. Burnett, p. 4.
68. Ibid., p. 139.
69. Carlyle, *On Heroes and Hero-Worship*, II, p. 1.
70. Tytler, p. 212.
71. Burnett, p. 2.
72. Ibid., p. 3.
73. Lieven, p. xvii.
74. Burnett, pp. 40–41.

75. Lawrence James argues in *The Aristocrats* (London: Abacus, 1009, 2epr. 2010) that 'the medieval aristocracy were always depicted as a physical elite and many were in life. A modern autopsy on the skeleton of Sir Bartholomew de Burghersh, who died in 1369 [...] revealed a sturdy man of nearly six foot with strong, muscular limbs. His physique was the result of regular exercise and a diet rich in protein [....] Knights were not just taller and stronger than their inferiors. Popular chivalric romances constantly drew attention to the fine features and fair complexions of noble heroes and heroines. Fiction often reflected reality' (p. 25).
76. Burnett, p. 5.
77. Ibid., p. 14.
78. Bryden, p. 80.
79. Burnett, p. 25.
80. Ibid.
81. As Tytler writes of physiognomy, 'The tendency to discover similarities between human beings and animals is, of course, practically as old as literature itself, though the animal comparison was, at least until the late eighteenth century, essentially metaphorical or symbolical [...] despite acknowledging the unity of nature and, hence, man's close kinship with animals, [the comparison] still held for the most part to the idea of the immutability of species as well as the unquestioned superiority of man in the chain of being' (250).
82. Ina Zweiniger-Bargielowska, *Managing the Body: Beauty, Health, and Fitness in Britain 1880–1939* (Oxford Scholarship Online, January 2011) (Abstract to 'Chapter 1'), https://doi.org/10.1093/acprof.oso/9780199280520.001.0001.
83. Burnett, p. 132.
84. Ibid., p. 220.
85. Ibid., p. 198; p. 230.
86. Carlyle, *On Heroes and Hero-Worship*, II, p. 108; Frazer, p. 108.

Works Cited

Anonymous 'The Origin of Man', *The Bristol Mercury and Daily Post* (5 April 1879), 9637, p.6.

Anonymous, 'Review of *The Prisoner of Zenda*', *Saturday Review of Politics, Literature, Science and Art*, 77:2013, 26 May 1894, pp. 556–57.

Beer, Gillian, *Darwin's Plots*, (London: Routledge & Kegan Paul, 1983).

Bryden, Inga, *Reinventing King Arthur* (Aldershot: Ashgate, 2005).

Burnett Tylor, Edward, *Primitive Culture* (London: J. Murray, 1871).

Carlyle, Thomas, *On Heroes, Hero-Worship and the Heroic in History* (1841) 2 vols (London: Macmillan and Co., Limited, 1926).

Carlyle, Thomas. *Sartor Resartus* (1834), ed. by Kerry McSweeney and Peter Sabor (Oxford: Oxford University Press, 2008).

Daly, Nicholas, *Ruritania: A Cultural History, from* The Prisoner of Zenda *to* The Princess Diaries (Oxford: Oxford University Press, 2020).

Darwin, Charles, *The Descent of Man* (New York: D. Appleton and Company, 1871).

Darwin, Charles. *On the Origin of Species* (1859), ed. by Jim Endersby (Cambridge: Cambridge University Press, 2009).

Elias, Norbert, *The Civilizing Process* (1939), trans. by Edmund Jephcott (1978), ed. by Eric Dunning, Johan Goudsblom and Stephen Mennell (Oxford: Blackwell Publishers, 1994, repr. 2000).

Fay, Elizabeth, *Romantic Medievalism* (Houndmills, Basingstoke: Palgrave, 2002).

Francis, Mark, *Herbert Spencer and the Invention of Modern Life* (2007) (London: Routledge, 2014).

Frazer, James, *The Golden Bough*, 1922 edition (London: Penguin, 1996).

Galton, Francis, *Inquiries into Human Faculty* (1883) (New York; J.M. Dent, 1919).

Gilmartin, Sophie, *Ancestry and Narrative in Nineteenth-Century British Literature* (Cambridge: Cambridge University Press, 1998).

Goldsworthy, Vesna, *Inventing Ruritania* (New Haven: Yale University Press, 1998).

Hawkins, Mike, *Social Darwinism in European and American Thought 1860–1946* (Cambridge: Cambridge University Press, 1997).

Heffer, Simon, *Moral Desperado: A Life of Thomas Carlyle* (London: Weidenfeld and Nicolson, 1995).

Hodgson Burnett, Frances, *The Lost Prince* (1915) (London: Heinemann, 1966).

Holloway, Lorretta M. and Jennifer A. Palmgren, 'Introduction', in *Beyond Arthurian Romances: The Reach of Victorian Medievalism*, ed. by Lorretta M. Holloway and Jennifer A. Palmgren (Houndmills, Basingstoke: Palgrave Macmillan, 2005), pp. 1–8.

Hope, Anthony, *The Heart of Princess Osra* (London: Longmans, Green, and Co., 1896).

Hope, Anthony. *The Prisoner of Zenda* (1894), ed. by Tony Watkins (Oxford: Oxford University Press, 2009).

Hope, Anthony. *Rupert of Hentzau* (1898) (London: Penguin, 2008).

Hope, Anthony. *Sophy of Kravonia* (1906) (Memphis: General Books, 2012).

Howie Wylie, William, *Thomas Carlyle: the man and his books* (London: T. Fisher Unwin, 1909).

Hurley, Kelly, *The Gothic Body* (Cambridge: Cambridge University Press, 1996).

James, Lawrence, *The Aristocrats* (London: Abacus, 2009; repr. 2010).

McCutcheon, G.B. [George Barr], *Beverly of Graustark* (New York: Dodd, Mead & Company, 1906).

McCutcheon, G.B. *Graustark: The Story of a Love Behind a Throne* (Chicago: Herbert S. Stone and Company, 1901).

McCutcheon, G.B. *Truxton King* (New York: Dodd, Mead & Company, 1909).
Menikoff, Barry, 'Prayers at Sunset' in *Robert Louis Stevenson Reconsidered*, ed. by William B. Jones, Jr (London: McFarland & Company, 2002), pp. 209–12.
Morris, Kevin L., *The Image of the Middle Ages in Romantic and Victorian Literature* (London: Croom Helm, 1984).
Niemeyer, Paul J., 'The Royal Red-Headed Variant: *The Prisoner of Zenda* and the 1893 Heredity Debates', *College Literature* 41:1 (Winter 2015), pp. 112–38.
Nordau, Max, *Degeneration* (1892), 7th ed., translated from the 2nd ed. of the German work (New York: D. Appleton and Company, 1895).
Nyman, Jopi, *Under English Eyes* (Amsterdam: Rodopi B.V., 2000).
Pearl, Sharrona, *About Faces* (Cambridge, MA: Harvard University Press, 2010).
Q., 'Review of *Prince Otto*', *The Academy*, 1442, 23, December 1899, pp. 750 (p. 750).
Secord, James A., *Visions of Science* (Oxford: Oxford University Press, 2014).
Simmons, Clare A., *Reversing the Conquest: History and Myth in Nineteenth-Century British Literature* (New Brunswick, NJ: Rutgers University Press, 2006).
Spencer, Herbert, *The Principles of Biology* (London: Williams and Norgate, 1864–67), II.
Stevenson, Robert Louis, *Prince Otto* (1885) (London: Chatto & Windus, 1912).
Stevenson, Robert Louis. *The Strange Case of Dr Jekyll and Mr Hyde and Other Tales* (1886), ed. by Roger Luckhurst (Oxford: Oxford University Press, 2006; reiss. 2008).
Tholuck, August, 'Review of *On Heroes and Hero-Worship*' (March 1847), in *The Critical Response to Thomas Carlyle' Major Works*, ed. by D.J. Trela and Rodger L. Tarr (Greenwood Press: London, 1997), pp. 100–02.
Trela, D.J., and Rodger L. Tarr's *The Critical Response to Thomas Carlyle's Major Works* (London: Greenwood Press, 1997).
Tytler, Graeme, *Physiognomy in the European Novel* (Princeton: Princeton University Press, 1982).
Vanden Bossche, Chris R., *Carlyle and the Search for Authority* (Columbus, OH: Ohio State University Press, 1991).
Wallace, Raymond, 'Cardboard Kingdoms', *San Jose Studies*, 13:2 (1987), pp. 23–26.
Warwick, Eden [George Jabet], *Notes on Noses* (1848), 2nd ed. (London: Richard Bentley, 1864).
Zimmerman, Virginia, *Excavating Victorians* (Albany: State University of New York Press, 2008).
Zweiniger-Bargielowska, Ina, *Managing the Body: Beauty, Health, and Fitness in Britain 1880–1939* (Oxford Scholarship Online, January 2011), https://doi.org/10.1093/acprof.oso/9780199280520.001.0001.

CHAPTER 6

'Nature Works On': Class Hierarchies in the Evolutionary Feudal

Introduction

In complete opposition to the chivalry and sprezzatura considered desirable in the aristocrats of Ruritanian Medievalism, the Medieval Revival also produced literature which contemplated a darker and perhaps more realistic form of medieval aristocracy—that which I have called the Evolutionary Feudal. Despite forming a clear cohort and having what I find to be more tightly defined characteristics than something like sensation fiction, at the end of the nineteenth-century Evolutionary Feudal texts were not grouped together into a defined, named genre. Instead, they span several literary and marketing categories, such as 'scientific romances', historical fiction, disaster fiction, horror, adventure, or even nature and travel writing. Often taking place in a futuristic, post-apocalyptic setting, the Evolutionary Feudal opposes the Chivalric Feudal of Ruritania, in that the Evolutionary draws heavily on natural history and Darwinian theory to depict aristocratic bodies, both of which Ruritania utterly ignores, challenges, or skews. The Evolutionary Feudal's views on aristocracy are as practical as Ruritania's views are fanciful. Though both genres are products of the Victorian Medieval Revival, they approach Medievalism from radically different angles. Where Ruritanian fiction hearkens back to the late medieval and early modern periods, calling on the high-romance traditions from the twelfth to the fifteenth century, the Evolutionary

A. Boucher, *Science, Medicine, and Aristocratic Lineage in Victorian Popular Fiction*, Palgrave Studies in Literature, Science and Medicine, https://doi.org/10.1007/978-3-031-41141-0_6

Feudal delves deeper into Western history. Despite often being set in the future, the class structures and aristocratic portrayals in these texts refer, instead, to a dark age or even prehistoric setting, where 'aristocracy' is often depicted as a tribal chieftainship or the alpha-dominance of the animal world.

Even where an Evolutionary Feudal text does not interact directly with Darwin (although most do), the genre as a whole is built on addressing questions regarding human origin and its fate in the natural world—questions which had been percolating in the British consciousness since at least the eighteenth century and had come to the forefront during the nineteenth-century *fin de siècle*.[1] As with George Eliot's ambitious, progressive Dr Lydgate in *Middlemarch*, there was a desire to source humanity's literal and metaphorical 'primitive tissue', to understand nature's relationship to current civilisations and current bodies.[2] The biologist T.H. Huxley, famously nicknamed 'Darwin's Bulldog', wrote in this period in his 'On the Method of Zadig' (1880) that science must be a field of retrospective divination: that science requires us to look back in order to develop predictive models of the future.[3] Concepts of 'society' and 'progress' were no longer givens. Even the socialite and muse Evelyn Nesbit tellingly writes of her days in *fin de siècle* high society, 'Nature is very cruel [...] and if civilization has overlaid us with delicacies and refinements, nature works on just as though social laws had no existence.'[4] In tracing humanity's roots back to its origin, Evolutionary Feudal texts address that most aspects of modern society—even something as elevated as the concept of the 'divine aristocracy' in Ruritanian fiction—were actually relics of a more barbarous time, and could perhaps lead to that barbarous time again. The organised, caring universe found in Ruritania is replaced by a harsh and indifferent system of nature; and while nature is structured and its developments are not arbitrary, there is no higher power or greater organisation than it.

Much like with Ruritanian fiction, the Evolutionary Feudal was a popular genre, but one that has (at least in this period) fewer examples than some of the more saturated genres in previous chapters; as such, I will once again look at the three major texts which serve as hallmarks of the genre and which, happily, span the fin de siècle period. As these build on each other and evolve the genre, I will once again look at them sequentially, rather than more thematically. My three Evolutionary Feudal case studies are Richard Jefferies's *After London* (1885), H.G. Wells's *The Time Machine* (1895), and M.P. Shiel's *The Purple Cloud* (1901). All three overtly contrast the Ruritanian model of Medievalism, are popular and ground-breaking examples of their genre at *the fin de siècle* and early twentieth century, and provide a wide-enough literary range to illustrate the disparateness of that genre—which aligns perfectly with the onset of Ruritanian fiction's popularity. That the Medieval Revival bifurcated into two such divergent literary styles at the same time, both of them very popular, speaks not only to the magnitude of Victorian preoccupation with historicity and origin-seeking, but also to the Victorians' fascination with aristocracy and *its* origins, rooted somewhere in the scientific or the medical; if there is one quality both Ruritanian and Evolutionary Feudal texts share, beyond their allusions to Great Britain's pre-modern history, it is a commentary on class and expectations about the heredity and evolution of the aristocratic body. Evolutionary Feudal texts diverge from Ruritania in their view that the aristocratic body is a product of its natural environment, rather than a product of divine placement. Evolutionary Feudal texts suggest that societal expectations of that body are the direct result of evolutionary forces on that body. In short, expectations of the body are formed in accordance with how that body develops and performs in nature over long periods of time. The aristocratic body is no better or worse, no more virtuous or sinful, no more glamorous or commonplace, than a lower-class body. Bodies, and our expectations or readings of them, merely stem from the long effects of sexual and natural selection. These texts assert that appearance, heredity, and bodily performance in nature are, over time, coded in the cultural consciousness, often to the point where the origin of these cultural codings is lost. In opposition to the perfect, physiognomic morality of the Ruritanian aristocrat's body, which exists somehow outside of evolutionary forces and is divorced from time to a cartoonish degree, the bodies of the Evolutionary Feudal are all just that: merely bodies. Zimmerman says of Victorian attitudes of archaeology and its resonating reflection of human impermanence, that 'the proximity of human remains to extinct faunal remains made the implications of geology for humanity very clear: people and their cultures are no more resistant to the passage of time than are bivalves or dinosaurs'.[5] Zimmerman is, of course, working from Gillian Beer's reading of the nineteenth-century evolutionary theory in which Beer argues that 'Darwinian theory [...] suggested that man was not fully equipped to understand the history of life on earth and that he might not be central to that history'.[6] Beer, in turn, echoes Frank McConnell's work on Victorian readings of H.G. Wells, in which he writes, 'If everything can be explained as an accidental development of life evolved just to preserve its own blind struggle for existence—everything *including* humanity—then what do morality or civilization finally mean [...]?'[7] The Victorian search for human and societal origin is not only the search for a record of our change, or even for a forecast of our future, but also a search for meaning.

The Evolutionary Feudal both fundamentally opposes and mirrors its sister-genre, the Chivalric Feudal. Remarkably, Carlyle's *On Heroes and Hero-Worship* may still be used as a philosophical framework, for though Carlyle assures us that the true hero-aristocrat is divinely elected and that humanity will always need a leader, he also insists that the aristocratic institution is not infallible, that our need for a leader is an animal response, that only 'true' aristocrats are divinely appointed but their dynastic lines are not immune from either biological or social change. Carlyle states that when aristocrats become unsuitable 'there have to come revolutions then' and they must be replaced with new and real elite heroes.[8] Primogeniture had, since at least the time of Reynolds's writing, if not far before, come under political fire for its irrelevance as a feudal relic or received stringent defence as a longstanding and therefore beneficial way of life—Evelyn Cecil in his *Primogeniture: A Short History of its Developments in Various Countries, and its Practical Effects* (1895) rather thinly argues, as pointed out by a contemporary critic, that primogeniture is 'for the common good' in that it is the only remaining system that could possibly keep large landed proprieties united.[9] Arthur J Munby, writing in *Fraser's Magazine* in 1870, reports that a bill has been brought in '[y]ear after year [...] to abolish the law of primogeniture, but year after year it has been thrown out or set aside by measures more engrossing, if not always more important'.[10] Munby goes on to trace the long, complex history and motives of primogeniture inheritance, some of which would find more resonant root in the Evolutionary Feudal, a genre which depicts leadership and class systems as subject to all of the crude, practical, and haphazard measures of the animal world from which they ostensibly stem. Aristocrats become aristocrats out of a 'natural' aristocracy, in its most literal interpretation; the institution survives, or not, to the degree that aristocrats are the 'best', or, to place it in Darwinian and Spencerian terminology, 'the fittest'.

Evolutionary Feudal texts are by no means as homogenous a group as Ruritanian fiction is. Although all three texts place class structures in the hands of biological systems, *After London* favours evolution, optimism, and a cyclical pattern that is always forward-looking, even in the face of total societal setback; *The Time Machine* employs more pessimistic views of progress, focusing on the supposed inevitability of human degeneration, while *The Purple Cloud* takes a firmly ambiguous view. Where *After London* has a clean reset of human culture that always aims towards progress, *The Time Machine* depicts human development as a series of messy peaks and valleys: the apex of cultural and evolutionary achievement is only possible after a slow upward climb, and can only result in a slow downward spiral. *The Purple Cloud* lies somewhere between the two and leaves equal potential for humanity to rebuild itself, as in *After London*, or to languish and disappear, as in *The Time Machine*. Further, though all three texts tie the concept of class to that of human origin and development inside the animal kingdom, *After London* suggests that the shift from aristocratic hegemony to middle-class hegemony is an inevitability of societal evolution. *The Time Machine* views the notion of the middle class as false and its formation and influence as merely societal delusions: you are always either upper class or lower class, the Haves or the Have Nots. *The Purple Cloud* argues, perhaps most cynically or pessimistically of all, that class is a product of luck: nothing more than being the right body in the right environment at the right time.

After London

After London was written in 1885 by popular nature writer, Richard Jefferies, who work remains woefully underexplored in academia; criticism which examines his texts' relationship to class is almost non-existent.[11] Jefferies as an author is difficult to classify since his name, as contextualised by Jefferies's biographer, W.J. Keith, 'is often to be found on the periphery of the English literary scene in that indistinct no-man's-land that skirts the boundaries of creative literature, natural history, and rural sociology'.[12] Jefferies came to literary prominence as a nature essayist, but his short novel *After London* became one of his best known works of fiction—popular enough to be one of a few of his texts still in print today—and introduced his work to a wider demographic.[13] Jefferies's views on class are likewise difficult to classify, since he was notoriously nonpartisan.[14] Even his clear hypothesis in *After London* on the origin and, indeed, the evolutionary necessity of aristocracy in early human culture is undone by his depictions of aristocratic confusion and failure in later evolutionary stages of civilisation. To place *After London* in the greater context of Jefferies's oeuvre, the only role that class plays in his nature writing is to show all the facets of agricultural life, displaying even-handed portrayals of both large landowners and the lower-classes labourers. Two of his books, *The Gamekeeper at Home* (1878) and *The Amateur Poacher* (1879) are told from the points-of-view of the titular enemies: the servant of the aristocracy who protects their land and game, and the lower-class agrarian poacher who steals that game. Both texts avoid class or political commentary and stick to the logistics and experiences of spending one's life in nature.

After London—a real departure from his previous work—depicts the aftermath of an unnamed cataclysmic event which has transformed England of the future into England of the early Middle Ages. Aristocrats once again dominate disconnected petty kingdoms on the back of serf labour. Much of the collective knowledge has been lost in the generations since the apocalypse, since 'the richer and upper classes made use of their money to escape [and t]hose left behind were mainly the lower and most ignorant'.[15] With no one left to understand it, all post-Renaissance technology has faded from memory; education of any sort is restricted to the upper classes; cities have fallen into disrepair and have been reclaimed by nature; many weaker species have become extinct and a man's social worth is in direct proportion to his physical strength. Nature and society have been reset, and Jefferies indicates that the development of a feudal aristocracy is society's primal setting, its roots coming from the animal kingdom.

The premise of *After London* works overtly in the confines of Darwin's insistence that natural selection may be best understood 'by taking the case of a country undergoing some physical change, for instance, of climate. The proportional numbers of its inhabitants would almost immediately undergo a change, and some species might become extinct.'[16] While Jefferies believed in evolution, and had certainly read and agreed with some of Darwin's theories, he was not a strict Darwinist.[17] Blomfield goes so far as to say that Jefferies 'promulgates anti-Darwinian discourses', and Peterson says 'his intention is anti-scientific (or a warning against too much science)', but these critiques cannot fully account for the mechanics seen in *After London*.[18] The novel's first section is revealingly titled 'The Relapse into Barbarism' and Jefferies spends nearly a quarter of the novel on it, discussing the status of the natural world: how the topography of England has changed, which plants have proved to be the most dominant, which animals have become extinct and, finally, a brief anthropological and epidemiological view of the structure of human life remaining. That so much emphasis is instantly placed on the natural world underscores Jefferies's constant message that all systems, be they human, plant, or animal, are products of nature. Importantly, the characters and plot are incidental to the greater story of Earth: both starting and ending the narrative *in medias res* to highlight the smallness of individual characters in the greater scheme of natural history, he argues that civilisations and social structures may forget the roots and paths of their evolution, but the act of forgetting does not negate the origin.

The reader enters human society a few generations after the apocalypse, and in that time it has just transitioned from animal packs and early-man tribes into organised classes, where Carlyle's system of the natural aristocrat is already on the wane through the introduction of primogeniture and its corresponding lack of meritocracy. Jefferies's pseudo-Dark Age setting is more than just recalling medieval period; it is the bridging state between the prehistoric and the modern. The narrative follows Felix Aquila, eldest son and heir to a minor baron. Through Felix we see the human element on the greater stage of nature and begin to understand the socio-Darwinian complexities of a culture in flux. At this stage of societal development, the aristocratic ideal is still (but only *just*) rooted in nature. Only a few generations previously, at the time of the apocalypse, the only aristocrat was a natural aristocrat. The wisest and strongest men were elected the leaders, and eventually 'assumed higher authority as the past was forgotten, and the original equality of all men lost in antiquity. The small enclosed farms of their fathers became enlarged to estates, the estates became towns, and thus, by degrees, the order of the nobility was formed.'[19] This definition of aristocracy is crucial to Jefferies's thesis: it answers the Victorian question of aristocratic origin and it speaks to the quality of human ego and how humanity will inevitably perceive itself as ascending over nature. The reader witnesses how a single man's power over the landscape can grow over time until he is no longer a component of or a resident on the land, but rather its owner who can force it to bend to his will—forgetting entirely that he and his descendants must bend to the will of nature in return.

This is not to say that Jefferies is, in any capacity, an anti-aristocracy reformer.[20] His texts depict feudalism as one stage of many in society's unstoppable evolution. In fact, he portrays aristocracy as a relatively positive and completely inescapable stage, in that it springs unbidden from the basic structures of the world; when the apocalypse returns humanity to barbarism, humanity retraces its exact footsteps in history. Keith writes of *After London* society, 'we see a new struggling civilization making the same tragic mistakes and blunders as the old. It is a vision (and this is crucial) not of evil but of ignorance.'[21] Jefferies's portrayed ignorance is, I argue, the ignorance of a child or adolescent; humanity has returned to an earlier stage of development, with feudal aristocracy being as a necessary phase in human social evolution. In a way, Jefferies is optimistic in his view of the human race, much like Carlyle, who says that the identification of a hero and the election of an aristocracy is inevitable: 'Hero-worship never dies, nor can die. Loyalty and Sovereignty are everlasting in the world.'[22] Jefferies sets his humans backwards at least a thousand years, and they carry on progressing and evolving as how he perceived nature would dictate. In his New Historicist reading of Jefferies's work, Brannigan writes, 'The crisis which haunts Jefferies [...] is of the imminent danger of society collapsing back into barbarism, and as such it shares its anxieties with other texts of the late Victorian era, most notably H.G. Wells's *The Time Machine*.'[23] While Jefferies was certainly ambivalent about the effects of technology, this text works largely against fears of human degeneration, in that his human characters seem so resistant to barbarism, at least in the long-term; the reader can witness *After London* society rebuilding itself.[24] *After London* reflects and undoes a common fear/comfort paradox which plagued Victorians as they gazed into the past: 'Faced with geological and archaeological ruin, nineteenth-century observers felt they witnessed at once the decay of the past and a preview of their own eventual ruin, yet paradoxically they also saw the persistence of the past, and therein lay hope for the future.'[25] Jefferies counters this 'decay of the past' through rhetoric which focuses on the endurance and reprises of the past. Zimmerman expounds that the history of nature shows the history of change as well as the history of repetitions or stagnations. Jefferies depicts mankind as physically and socially evolving, but in positive, predictable, and set ways. Jefferies traces the changes that have taken place in a few short generations and the social confusion and cultural ripples as humans move away from the roots of the aristocracy's origin towards more modern and middle-class values. Through his examination of class, heredity, and bodily expectations, he presents a clear linear path from where human thought and ideology have been, to where they are headed.

All societies, transhistorically, express preferences for and idealise certain physical and physiological traits. As has been amply discussed, this idealisation is especially true of the physicality of rulers, whose bodies either shape these preferences or are, to some extent, policed by others for its success or failure in adhering to that ideal. The Evolutionary Feudal adds a complicating factor by tracing the roots of this socially constructed process back to sexual selection and survival—or, rather, the texts illustrate how socially constructed processes around mating choice and bodily norms can be in outright conflict with more unmediated forms of sexual selection. Even the mechanisms by which mates are selected can change and evolve. Where in Ruritania the idea ruler was male, energetic, wiry, but restrained, well groomed, and temperate, the society of *After London* prefers an unevolved masculinity. Here, the ideal aristocratic man is brutish, animalistic, thickly muscled, and strangely hairy: hair becomes such a gauge of a man's worth that 'No slaves were allowed to wear the moustache', since they did not qualify as men.[26] This also speaks to society's progression from the natural aristocracy/natural slave ethos—it is not that slaves cannot grow moustaches but are rather not allowed to grow them. Masculinity becomes the sole property of the ruling class as cultural decisions take the place of natural properties and eventually replace them.

Felix does not conform to this ideal and is therefore an outsider, a shy scholar depicted as 'rather dainty', scorned by the other aggressively masculine alpha-males of his class.[27] Felix appears too fragile to survive in a harsh environment, too weak to participate in war, and too effeminate to produce children. When he is called 'so slender a stripling', it not only conjures up an unfortunate phallic allusion, but also to the idea of competition and survival of the fittest; Felix is a small tree in the shade of greater trees.[28] In Felix's younger brother, Oliver, we see *After London*'s ideal aristocratic manhood embodied:

> Oliver's whole delight was in exercise and sport. The boldest rider, the best swimmer, the best at leaping, at hurling the dart or the heavy hammer, ever ready for tilt or tournament, his whole life was spent with horse, sword, and lance. A year younger than Felix, he was at least ten years physically older [with] massive shoulders and immense arms, brown and hairy [...] every inch a natural king of men. That very physical preponderance and animal beauty was perhaps his bane, for his comrades were so many, and his love adventures so innumerable, that they left him no time for serious ambition.[29]

Oliver's conformity to the animal world, even down to his excessive hairiness, indicates his suitability to be a leader in the political reality of the text. That he is 'every inch a natural king of men' shows the priorities of the community, in which physical hardiness and obvious physical interests are common-sense prerequisites for leadership, in a time where war, famine, and predators are commonplace and there are little or no mechanised or technological solutions to compensate for physical shortcomings. The most intriguing dynamic, though, is Jefferies showing us a world in flux and how Oliver's supposedly unevolved, almost neolithic body has already been compromised by civilising forces.

Oliver does not actually lead or, in fact, do anything of value for his community, but his body is celebrated as both a figurehead of cherished ideals and in a minor capacity as a 'pet' put on display through competition, much like a prized horse or dog. Competition becomes the key word in this rhetoric: Oliver's tilts, tournaments, and feats of strength serve as an example of Darwinism becoming clouded by advancing society, turning survival of the fittest into a spectator sport to reassure those watching. Competition, in the sense that Oliver understands it, is still a crucial element of leadership in Oliver's world, but the meaning is gradually eroding. *After London* society understands Oliver's style of masculinity and leadership; it does not understand Felix's. Yet the reader is set up to prefer Felix, not only because he is the sympathetic protagonist or because he mirrors many of the Victorian readers' mores in an otherwise unfamiliar and severe world, but rather because Felix's more delicate but *adaptable* body is represented as the foil to tyranny, which appears to be rampant in a time of cultural evolution and confusion. Jefferies writes, 'The principal tyrant [i.e. the king] is supported by the nobles, that they in their turn may tyrannise over the merchants, and they again over all the workmen of their shops and bazaars.'[30] Oliver is depicted as a content member of this society from which he deeply benefits and in which he comfortably fits. Oliver's body—brutish but ultimately ornamental—is a necessary object to continue the chain of petty tyranny and an indicator of his stagnation: he is incapable of adapting either his increasingly irrelevant body type, has no strength of character to do so, nor any social imperative to want to break the cycle. The conditions in which humans exist in Jefferies's world are changing; the qualities valued in selection of mate will therefore follow suit.

Tensions rise to the surface, Jefferies illustrates, because this society has already developed a sense of lineage and inheritance but is no longer able to rely on a 'natural' aristocracy. Men were initially selected for leadership roles because their traits calibrated the whole aristocratic institution to the specific challenges of their world. The faulty assumption of primogeniture is that those traits will either be duplicated perfectly down the line without dilution (*à la* Ruritanian heredity) or that those traits will still be the most useful—that heredity and the world can both be frozen in amber. The

England of *After London* has strong primogeniture laws, and Felix, as the eldest son, is therefore an unwanted, unsuitable, and yet inescapable leader. Indeed, Jefferies, Carlyle, and Darwin all agree that the socio-physical excellence of one generation is not hereditarily guaranteed in the next, and that rule by primogeniture could be a doomed practice, subjected to endless rebellions by its own faulty logic. Primogeniture and systems of legal inheritance had been highly and publicly debated in Great Britain since at least the 1820s and were still being debated by the time of *After London*'s publication. In 1837, an anonymous author in *Tait's Edinburgh Magazine* wrote in presciently Jefferies-esque terms, 'The tendency of primogeniture, like injustice of all kinds, is to check the wholesome and natural competition of individuals for pre-eminence and station, and to convert society from a clear stream of running water, to a stagnant pool.'[31] Darwin echoes these sentiments from a more clinical vantage:

> Primogeniture with entailed estates is a more direct evil [for Natural Selection], though it may formerly have been a great advantage by the creation of a dominant class [...] The eldest sons, though they may be weak in body or mind, generally marry, while younger sons, however superior in these respects, do not so generally marry. Nor can worthless eldest sons with entailed estates squander their wealth [and thereby leave power].[32]

Jefferies, writing with more neutrality, says of the grand Prince who rules the territory in which Felix resides, '[he] was not a cruel man, nor a benevolent, neither clever nor foolish, neither strong nor weak; simply an ordinary, a very ordinary being, who chanced to sit upon a throne because his ancestors did.'[33] In *After London*, the disparateness of who is able to rule and who has the legal right to rule comes to the foreground as crucial step in society and perhaps even human evolution, a step in which natural and social laws begin to separate:

> As they [nobles] intermarried only among themselves, they preserved a certain individuality. At this day a noble is at once known, no matter how coarsely he may be dressed, or how brutal his habits, by his delicacy of feature, his air of command, even by his softness of skin and fineness of hair.[34]

That 'brutal habits' are suddenly the antithesis of what an aristocrat *should* be indicates the trend is heading away from brutality instead of towards it—that the vigour and roughness which aristocrats like Oliver hoped to perpetuate through endogamy are the very things to be lost by it. Jefferies also undoes the notion, which was so prevalent in Ruritanian fiction, that an aristocrat's body is unmistakably classed. In Ruritania, these cues and features are God-given; the body becomes coded by a higher power to assert its right to rule on earth and to illuminate the highest physiognomic complexities. Jefferies does not deny that aristocratic demographics might be recognisable, but he takes their physical conspicuousness from the hands of God and returns it to nature: it is mere inbreeding that brings about such homogeneity. Not only does primogeniture therefore keep the 'Ablest' man from moving upwards to a place of leadership, but it also recycles negative characteristics through heredity in a closed-off and exclusive group, which Darwin argues, in almost every instance in every species, 'diminishes vigour and fertility'.[35] This will not stop masculine hardiness from being the favoured bodily ideal of the aristocracy, but it will mean that this expectation will be disappointed more and more often, to the point where the expectation of frailty and physical failure becomes the norm.

Through the cultural evolution of these bodily codes, one can see how society has gone from the glorification of Oliver's body to the glorification of Rudolf's in *The Prisoner of Zenda*. After enough frustrated attempts, a new definition of the ideal body must be reached; the Victorian aristocracy might never again produce a brutal Oliver warlord, but it *could* produce the gentlemanly athleticism of the chivalric Rudolf, or the inventive and determined Felix. Jefferies excavates the historical basis of social expectation of aristocrats—men in particular—and leads it to its present state.

If, as Jefferies proposes, the development of an aristocracy is society's natural, necessary adolescent setting but primogeniture complicates this institution to the point where it will not work, Jefferies considers the natural 'adulthood' of society to be the development of the middle class. Felix is unambiguously coded as a Victorian middle-class figure: his primary struggle in borne from a desire for mobility: 'As men were born so they lived; they could not advance.'[36] Perhaps more accurately than 'mobility', Felix seeks a form of success that can exist somewhere between the realm of the serfs and the aristocrats—particularly, his dream of middle-class domesticity. Felix wants to marry Aurora, the much-coveted daughter of a powerful noble, and settle with her into a quiet, private domestic space. Aurora seemingly senses that Felix has appropriate (read: Victorian) traits to pass to their children and selects him above other men, for he complements her own evolutionary stage; she also covets the quiet domestic space and the intellectual study which Jefferies depicts as at odds with feudal aristocratic duties and modes.[37] Aurora's choice of mate is complexly negated by her father, who finds Felix to be an inappropriate mate for his daughter: heir or not, Felix does not possess the requisite aristocratic qualities. In doing so, Aurora's father perhaps indicates his own lack of synchronisation with cultural evolution: he is either deeply out-dated and conforms to the earliest incarnations of aristocracy, acquired through physical merit and no longer particularly useful in their shifting society, or, like Felix and Aurora, he has progressed further than the rest of society at its current stage, and sees little value in the principle of primogeniture for its own sake.

To prove his worth, Felix decides to travel through unmapped territory and found his own castle-city over which he can rule, and to which he can bring Aurora; here we see the transition of ideology from aristocratic mores to middle-class mores—Felix, who is legally entitled to inherit an estate prefers, instead, to earn his own. The meritocracy of the middle class reinvigorates the 'natural' aristocracy and a return to 'survival of the fittest', supposedly without any manufactured restraints. As Felix carves his own boat to set forth on his adventure, Jefferies writes, 'He could easily have ordered half-a-dozen men to throw the tree, and they would have obeyed immediately; but [u]nless he did it himself its importance and value to him would have been diminished.'[38] This is a clear rejection of the power to which he is entitled as the son of a baron, and is Felix's personal Darwinian test to see if he is worthy both of Aurora and of survival. He even intends to present Aurora with some 'peacock's feathers [which are] rare and difficult to get' in order to best attract her.[39] The bright peacock feathers used by humans for ornamentation are strictly from male peacocks, used primarily for courting—depicted by Darwin as a perfect example of sexual selection—and Felix plans to use the feathers as a surrogate body part to help him stand above other men, though the prized body part is not *his*.[40] He is presenting, instead, his own middle-class ingenuity at procuring a rarity, and thus the Darwinian adaptability, flexibility, and competitive manoeuvring which can only exist outside of the limiting feudal system. Felix's brief adventure is cut short when he founds his own city and sets back home to claim Aurora; his success or failure in this final venture is unknown. While initially an unsatisfying ending, it is but one further way in which Jefferies inculcates the smallness of mankind. Felix's personal growth is irrelevant—if his mission is stymied, nature will carry on without him; the middle classes have begun.

The Time Machine

While it is hardly ground-breaking to view H.G. Wells's 1895 novel *The Time Machine* as a commentary on social division, no exploration of Darwinist or class-conflict literature would be complete without it. The novel tells the story of the Time Traveller, an unnamed, prosperous Victorian scientist, who invents a time machine and travels forward to the year 802,701, where a vague, post-apocalyptic dark age has reclaimed England. He observes, to his astonishment, that the human race has degenerated into two separate species: the weak, beautiful, surface-dwelling Eloi and the brutish, servile, subterranean Morlocks. England of the future initially appears to be a utopian feudalism, in which all conflict and hardship have been eradicated and merry lords are served by satisfied, or even complicit, servants. In reality, the relationship between the two classes is more akin to *After London* than Ruritania, fuelled by a grim, nature-driven symbiosis where each group fulfils a survival need for the other. The Time Traveller returns to the nineteenth century to tell his story, and eventually travels forward even further in time to watch the end of Earth; he is never heard from in his own age again.

Unlike the narration of *After London*, which contains a structural optimism about the future of humanity and the resilience of the middle classes, the narrative structure of *The Time Machine* reinforces Wells's pessimism. Where the unknown narrator of *After London* could only look backwards upon human history and seemed to do so with relief at evolving out of its earliest forms, the narrators of *The Time Machine* have the dubious luxury of looking both backwards and forwards in time. Where *After London*'s narrator could infer that feudalism was a necessary, if bleak, steppingstone to a potentially brighter future, the Time Traveller knows that his own time was the apex of civilisation, leading to an assuredly darker future. Although the Time Traveller tells the story, it is given to the reader by an intermediary, retold to us by a guest at the Time Traveller's home. The future is seen through three sets of eyes, the Time Traveller, his guest, and then the reader, with each relay accumulating the anxiety of the others. The string of narration serves to mimic time itself: the reader is divorced from first-hand knowledge of the future by two degrees of narration, and yet is still connected to it. The knowledge of what is to come, without direct experience of it, only serves to heighten anxiety Darwinian and (d) evolutionary modes of thought. Gillian Beer connects literary form to evolutionary content by arguing that '[b]ecause of its preoccupation with time and with change evolutionary theory has inherent affinities with the problems and process of narrative'.[41] The relay-narration utilised in *The Time Machine*, which is both linear and non-linear, can therefore be read as a clear manifestation of *fin de siècle* anxieties in literary form.

Both Wells and his text take manifestly pro-Darwinian stances. Wells held an early degree in zoology and was trained by biologist T.H. Huxley (mentioned above as 'Darwin's bulldog'), whom Wells deeply admired.[42] While it is difficult to deny or ignore Wells's adherence to the Darwinian model of evolution, his application of that model to class schisms is a far richer and more nebulous area of examination.[43] Wells had a complicated relationship with class: his mother was a lady's maid, who considered herself socially superior to Wells's gardener-turned-shopkeeper father, and brought up Wells with the hopes of his becoming a gentleman.[44] Class-consciousness was as a prominent part of his childhood as Darwinian theory was a part of his young adulthood; both would continuously inform his adult life and his writing, in particular his preoccupation with how class tensions would play out on a long-term evolutionary basis.[45]

> *The Time Machine* can be read, as we shall see, as a prophecy of the effects of rampant industrialization on that class conflict which was already, in the nineteenth century, a social powder keg. Disraeli had warned—and Marx had demonstrated—that the industrialized state was in danger of becoming two nations, the rich and the poor; but the real horror, Wells warns, is that they might become two races, mutually uncomprehending and murderously divided.[46]

Critics disagree on the precise conceptual split in Wells's Eloi and Morlocks. Some read the division as one between the Aesthetes and the Utilitarians.[47] Others believe it is the upper class splitting from the middle and labouring classes, or the upper and middle classes splitting from the labouring classes, or rural-dwellers splitting from the city-dwellers, or the Communists splitting from the Capitalists.[48] It is unlikely that even Wells himself had defined this split absolutely: 'in an earlier version of *The Time Machine* the Eloi were descendants of 1890s aesthetes, and the Morlocks were the descendants of the aesthetes' natural enemies, the middle class materialists', but Wells eradicates this distinction in later drafts, purposely making the social origins of each group more ambiguous.[49]

Wells's one clear demarcation of the split, as voiced by the Time Traveller, states: 'above ground you must have the Haves, pursuing pleasure and comfort and beauty, and below ground the Have-nots, the Workers getting continually adapted to the conditions of their labour.'[50] From this statement and from his absolute social and biological alienation of the 'Haves' from the 'Have-nots', Wells's text argues a Darwinian discourse in direct opposition to Jefferies: instead of theorising that feudalism is mankind's embryonic setting and that rise of the middle class is an evolutionary inevitability, Wells depicts feudalism as mankind's *only* setting, that the development of a middle class is a delusion, and that this stagnation of the class system will lead to human degeneration. In Wells's model, 'Having' or 'Having-not' are inheritable traits subject to the ravages of time and evolution; just as with the 'fitness' of bodies leading to primogeniture, these traits have social roots so long that they are often forgotten by modern society. Wells confounds the Victorian preoccupation with time by construing to his readers that the past never leaves, it only evolves; the Victorian exploration of human antecedents and genesis is ultimately as futile as the Time Traveller's trip into the future. The comfortable middle-class background from which the Time Traveller originates is a camouflaged medieval feudalism as extreme as that of the Eloi and Morlocks although he is trained by familiarity not to recognise it as such. The middle classes operate as a smokescreen for a feudal system which was never truly eradicated, with hegemonic parameters merely expanding to include more members, where 'affluence' replaced 'titles' as an entrance prerequisite. This new definition may not be as devastating to the elitist boundaries of the traditional upper class as one may think—and, indeed, this is what some silver fork novels argued more than half a century earlier. Affluence is still proportionally rare, highly inheritable, and can produce a level of cultural and genetic homogeny through its intermarriage of members. Even without title, affluence still enables lives of leisure, where lack of hardship makes room for a preoccupation with the aesthetics, where material goods are readily available, and where vast quantities of land are reserved for them alone. These are the qualifiers of Wells's slightly more inclusive reality of aristocracy.

Even the name 'Eloi' 'carries several obvious associations, suggesting not only their *elfin* looks, but also *èloignè*, and their apparent status as an *èlite*'.[51] Despite the Time Traveller's initial view of the Eloi as a distinctly alien race, his quick acclimation reveals a distinct kinship. Martin Willis reads the Time Traveller himself as a more traditionally aristocratic figure than even a middle-class one, whose laboratory conditions and scientific endeavours are tied to older models of science as a private, gentlemanly hobby; he is not, Willis argues, operating in the more modern and professionalised laboratory or clinical contexts of Wells's later titular characters, Dr Moreau and Griffin, the Invisible Man.[52] The Eloi's physical bodies are aristocratic to the point of caricature, hearkening back to the same tropes and physical qualifiers of the nobility that were examined in previous chapters. The Time Traveller says:

> I saw a group of figures clad in rich soft robes [...] One of these emerged in a pathway [...] He was a slight creature—perhaps four feet high—clad in a purple tunic, girdled at the waist with a leather belt [...] He struck me as being a very beautiful and graceful creature, but indescribably frail. His flushed faced reminded me of the more beautiful kind of consumptive—that hectic beauty of which we used to hear so much. At the sight of him I suddenly regained my confidence.[53]

That the male Eloi is graceful and frail, which are words often reserved in literature to describe female beauty, looks back to gender and health critiques of the aristocracy in previous chapters; Wells disorients the linear structure of time by reaching into society's past and projecting its representations of aristocrats onto the literary present which depicts the ultimate future. These gender issues are only heightened by the Time Traveller's later assertion that the Eloi 'all had the same form of costume, the same soft hairless visage, and the same girlish rotundity of limb' and that, due to their leisurely lifestyle, the 'the specialization of the sexes with reference to their children's needs disappears'.[54] The effeminacy of the Eloi is not used in this context specifically to denote any notion of homosexuality, but rather 'in its older, traditional sense, to refer to a male person or institution weakened by luxury or inactivity'.[55] Not only are the Eloi still the effete nobles of the past, but they are also stuck in a physical prepubescence where the distinguishing characteristics of each sex are not immediately apparent. The 'Haves' have become so inured to comfort, while wealth and primogeniture have so eradicated all biological need for 'survival of the fittest' in favour of pure socioeconomic considerations, that only feeble, effete children remain of the upper and middle classes. Darwinist mechanisms in human reproduction are themselves being eradicated—or at the very least polluted through too strong socioeconomic divides and class framings.

The focus on the Elois' bodies is a further signifier of their aristocratic status. In fact, the Time Traveller insists upon narrating at length about their costumes and physicality, using the lexis of their bodies (at least as he comprehends and interprets them) as the first indicator of the *time and place* in which he has landed, tying the Eloi to the euchronic setting as firmly as any aristocrat in Ruritania. The Morlocks, on the other hand, are hardly glimpsed and the little that is seen of them is deemed to be so hideous and inhuman by the Time Traveller that it is better if one does not see them at all. A Morlock is a 'bleached, obscene, nocturnal Thing', an underground species, not worthy of attention or vision; their purpose is functional, not decorative or interpretive. It is not surprising when the Time Traveller infers, based solely on his contemporary judgements of class appearance, that the Morlocks are the servants of the Eloi, who

> might once have been the favoured aristocracy, and the Morlocks their mechanical servants [...] The Eloi, like the Carlovingian kings, had decayed to a mere beautiful futility [... T]he Morlocks made their garments, I inferred, and maintained them in their habitual needs, perhaps through the survival of an old habit of service.[56]

What the Time Traveller is never able to do is consciously connect the Morlocks to his own servants, largely because of the servants' functionality has already rendered them invisible. Despite the Time Traveller maintaining a wealthy home, being seen to ring the bell for servants, and hosting a large party for his friends, neither he nor the reader is ever alerted to a servant's presence. His wishes are obeyed, and his food served by the same invisible, subterranean force. The sole service-workers that the Time Traveller addresses are Mrs Watchett and Hillyer, presumably his housekeeper and butler or valet, who are both just highly enough ranked to merit names and bodies and whom the Time Traveller only notices *after* his class-riddled adventure with the Eloi and Morlocks, and only very briefly. As he returns to his own time, he sees these two servants speed past him as time rewinds, as though these Morlock predecessors regress in real time and return to just enough humanity to render them visible; it is presumably already too late for his other, lower-ranking servants.

While the Morlocks seem hardwired to serve the Eloi out of thousands of years of habit—perhaps winking slightly at a Lamarckian idea of the inheritance of acquired characteristics, discussed more prominently in the chapter on sensation fiction—the Morlocks' long-rooted behaviour has evolved a secondary (or perhaps primary) purpose: they tend to the Eloi so they may eat them. 'These Eloi were mere fatted cattle, which the ant-like Morlocks preserved and preyed upon—probably saw to the breeding of.'[57] Through this revelation, Wells combines class tension and degenerative tension in a single horrifying moment in which the Time Traveller first considers civilisation to be fully and abhorrently collapsed; Norbert Elias convincingly argues in *The Civilizing Process* that one of the most defining characteristics of any society is in its people's attitudes towards meat eating.[58] The labourers of the nineteenth century, the Time Traveller realises, were so dehumanised through structural inequality that their fate is to turn feral and return to humanity's animalistic origins, akin to 'our cannibal ancestors of three or four thousand years ago', where they lose any moralistic qualms about consuming their sister-species.[59] The eating of the Eloi is a pragmatic practice, if also possibly an ideological one; there is little other food present, the Morlocks grow increasingly feral as they serve the Eloi, and the Eloi grow weaker from this feudalism until they no longer possess strength or survival instinct, adapting into the perfect prey. The Time Traveller says of them, 'I never met people more indolent or more easily fatigued.'[60] That the Eloi provide food for the Morlocks is a twofold parody class structures: the first being an inversion of the upper-class consumption of working-class bodies through the iniquities of both feudalism and capitalism; the second, meanwhile, nods to the material welfare that aristocrats supposedly gave to their subjects in return for service, and the system in which an aristocratic body was tied to the land to serve as an idol or scapegoat in respective times of feast or famine.

The Time Traveller's supposition that the Morlocks see 'to the breeding of' the Eloi further exemplifies his naiveté of the past, present and future, while reinforcing Wells's own (perhaps mis-)understanding of the Darwinian model of sexual selection. The Time Traveller's rash judgements about the future, which he often later admits to be wrong ('This, I must remind you, was my speculation at the time. Later, I was to appreciate how far it fell short of the reality'), trains the reader to be suspicious of the Traveller's instant and uninformed conjectures.[61] In this instance, the Time Traveller's narrative has already disproved own his belief that the Morlocks were a sort of eugenicist group that consciously organised the Eloi's breeding practices. The Time Traveller states several chapters before his discovery of the Morlock's cannibalism that the Eloi 'spent all their time in playing gently, in bathing in the river, [and] in making love in a half-playful fashion', and that their social and recreational practices, as far as he observes, are in no way impeded by the Morlocks who prefer predatory stealth and subterranean safety over explicit surface-power.[62]

Wells sheds further, though not explicit, social commentary on the Eloi's feeble physicality and how the system has spiralled into an evolutionary self-destruction. Representations of the Eloi illustrate how aristocratic bodily ideals change as an aristocrat's evolutionary purpose alters. In the case of the Eloi, a paradigm shift in physical preference happened somewhere after their species-wide enervation and the Morlock's modification into cannibals. Where they may once have been Jefferies's natural aristocrat, whose purpose was to survive conflict and hardship for the people, or Ruritania's chivalric aristocrat whose purpose was to be a divine figurehead for the people, now an aristocrat's purpose is to remain lazy and grow just large enough to be *eaten* by the people. Darkly, in all three of these fictional models, the unifying 'essence' of an aristocrat is for them to be somehow bodily consumed by and for the classes they rule.

It logically follows that if all Eloi are easy prey and there is no need to pick off the weakest members, then a hunting Morlock will choose the most tempting meal, that is, the Eloi with the most meat on their body. In terms of sexual selection, the Eloi females should therefore choose weakest and sickliest males to father children, in the hope of making offspring as un-tempting as possible for the Morlocks and, perversely, ensure their offspring's greatest chance of survival in this odd predatory context, through its dubious chance of survival in a *normal* context: 'that individuals having any advantage, however slight, over others, would have the best chance of surviving and of procreating their kind' and 'those which are best fitted for their places in nature, will leave most progeny'.[63]

With the Eloi, bodily ideals have gone from 'survival of the fittest' to 'survival of the least fit', which will perpetuate a system of increasingly weakening aristocrats and toughening labourers until one can no longer sustain the other, and both species die out entirely. The Time Traveller, as the reader learns, gives the Morlocks far too much credit in assuming that the Eloi's breeding pattern is a conscious decision and calculated effort on the part of the predators. Nature, not the Morlocks, is the ultimate clinical organiser of evolutionary systems, and Wells portrays that the one constant trait of humanity is its animalistic desire for preservation, even at their own long-term expense.

The Purple Cloud

It is fitting, given the perspective of the genre, that Ruritanian fiction showed little evolution in style, content, and themes over the decades of its popularity; is it equally appropriate that the Evolutionary Feudal evolved a great deal as a genre, and still continues to do so. M.P. Shiel's Decadent post-apocalyptic novel *The Purple Cloud* works heavily from both Jefferies's and Wells's texts while bringing the Evolutionary Feudal genre fully into a new era.[64] It is stylistically more experimental, more explicit about matters of violence and sexuality (which are so crucial to the spirit of the genre and to popular and scientific concepts of survival and competition), and perhaps even bleaker in outlook about the future of the human race than *The Time Machine*.[65] Its timely publication coincided with the death of Queen Victoria, a situation which lent itself neatly and with even more resonance to the genre's themes of ambiguity about social identities that were formed well in the past and confusion over what the future held in store.[66] In his work on nineteenth-century apocalypse fiction set in the Thames Valley, Patrick Parrinder links *The Purple Cloud* with *After London* through what he considers to be their mutual revelry 'in the destruction of civilization and the opportunity it provides for a return to an idyllic, barbaric existence'.[67] Although both texts certainly portray the barbarity of a pseudo-medieval landscape, the novels' perspectives on returning to such origins may be far less idyllic and far more ambiguous than Parrinder argues, especially where it concerns theorising about the scientific history of the aristocracy.

In addition to redrawing boundaries for the Evolutionary Feudal, Shiel also constructs a new paradigm about the Darwinian history of the aristocratic body. This paradigm, unlike the ones seen in *After London* and *The Time Machine* is not necessarily reliant upon 'natural aristocrat' ancestors or the degenerative forces of too much 'Having' and not enough competition. Rather, for Shiel, aristocrats are made so through blind luck, with small considerations being paid to heartiness and innovation—and certainly no consideration being paid to Ruritanian or Carlylean morality. The aristocratic body is a body in the right place, at the right time, and it is through random good fortune that aristocracies are formed in what is probably an uncaring Darwinian universe.

Published in 1901, the narrative of *The Purple Cloud* is relayed to the audience by a seer in a psychical trance. Whether a true account of things to come in the near future, or the true record of things that have passed, or just a story, the seer does not know. The protagonist and narrator of her tale is Adam Jeffson, a doctor with high society connections through his engagement to a countess, Clodaugh. The world is in raptures over the potential discovery of the North Pole, leading an American millionaire to award a fortune to the first man to set foot on the Pole. When an expedition is planned by some men of Jeffson's acquaintance, Countess Clodaugh discovers that her nephew is one of their number. She poisons him so Jeffson may fill his place, win the fortune, and thus become her social equal.

After many months at sea and much violence amongst his crewmates (all of whom wish to be the first man at the Pole and are willing to kill the others to obtain the fortune), only Jeffson remains alive. The exact cause is ambiguous, but Jeffson reaching the Pole coincides with a noxious, quick-travelling purple cloud being unleashed upon the globe. Shiel does not specify whether the cloud is the Earth's physical reaction to Jeffson's presence at the Pole, a divine punishment for man's curiosity, or merely coincidental. Following in the cloud's wake, Jeffson discovers the death of all human and animal life on the planet. Jeffson grows mad in his solitude, declares himself emperor of the world ('I am ten times crowned Lord and Emperor; I sit a hundred times enthroned in confirmed, obese old Majesty'), and spends the next decades alternating between building a magnificent palace to his own glory ('the only human palace worthy [of] the King of Earth'), and sailing the world in order to burn every city to the ground to destroy any remnants of civilisation ('In spite of the restraining influence of this palace, I have burned and burned. I have burned two hundred cities and countrysides [*sic*]').[68] The world returns to a dark age of pillage and terror through the rapid devolution and destruction of human endeavour, which is only interrupted by the reverent and superstitious glorification of an autocrat. Jeffson discovers in his travels a feral young woman, still alive by chance, whom he domesticates and keeps as a pet. They eventually fall in love and presumably go on to repopulate Earth.

As with many texts before it, *The Purple Cloud* uses the ambiguous provenance of the aristocrat to interrogate ideas around lineage, mental health, and criminality, and the racial or geographical elements which are often included in theories of heritage, evolution, and degeneracy. Shiel—who was born in the British West Indies, who was bizarrely crowned 'king' of an uninhabitable islet near Montserrat by his own father, and who was himself possibly of mixed race—has sparked in critics abundant speculation over the nature of the cringe-worthy delusions of grandeur, sexism, racism, and anti-Semitism seen in much of his writing.[69] In *The Purple Cloud* alone, the mad protagonist models his kingship and appearance on those whom he perceives to be despots of the Continent or the East, including Nero and Nebuchadnezzar, saying:

> My body has fattened, and my girth now fills out to a portly roundness its broad Babylonish [*sic*] girdle of crimson cloth, minutely gold-embroidered, and hung with silver, copper and gold coins of the Orient; my beard, still black, sweeps in two divergent sheaves to my hips, flustered by every wind; as I walk through this palace, the amber-and-silver floor reflects in its depths my low-necked, short-armed robe of purple, blue, and scarlet, a-glow with luminous stones.[70]

As discussed in the chapter on sensation fiction, the connection between racial or national identity and degeneracy has a long and full history, which will be alluded to in brief; these topics are far more robustly covered elsewhere will largely go beyond the remit of this chapter. *The Purple Cloud* is overstuffed with a tangle of anxieties and, for the purposes of scope, references to Jeffson's newly declared aristocratic body will focus on issues of class, history, lineage, and Darwinism, rather than his equally prominent but far more muddled stances on racial performance.

The same goes for the novel's great many theological questions, which often go hand-in-hand with the questions upon which Evolutionary Feudal texts are often based: those of class, identity, history, and human purpose. Religion in *The Purple Cloud* is so overt as to be almost parodic: Adam Jeffson believes his life to be guided and wrestled over by a 'White Power' and a 'Black Power', racially loaded terms which act as the clichéd angel and devil on each shoulder. When Jeffson finally attempts to name his feral companion, she (who by this point has learned to read the Bible), suggests 'Eve'. Jeffson responds, 'not Eve, anything but that: for *my* name is Adam, and if I called you Eve, that would be simply absurd, and we do not want to be ridiculous'.[71] They eventually settle on the name 'Leda' which, of course, has its own associations with deity. The issues of religion are simply too numerous and expansive to be considered fully here.

Finally, it must be noted that *The Purple Cloud* differs from the other two Evolutionary Feudal texts in that the pseudo-medieval landscape of this novel eventually expands beyond England, while *After London* and *The Time Machine* remain firmly inside its borders. This expansion serves three purposes. Firstly, it expresses the universality of both aristocracy and Darwinism in all societies and cultures. Secondly, it speaks to the globalisation of wealth, power, and empire at the turn of the century. Thirdly, it is a necessary means of creating a Biblical allusion: if all humanity on Earth derived and expanded from a single couple, from where does aristocracy originate? And, conversely, if all humanity on Earth dwindles back down to a single couple, where does aristocracy go? Working from the 'Polarion' theory of nineteenth-century occultist, medium, and author Helena Blavatsky, who posited that the North Pole is a sacred land from which 'our ancestral "Root-Race" originated', Shiel explores the dangers and desires of exploring human origin, especially on a global scale.[72] Shiel's decision to take the action out of England opens up the potential to address new concerns about aristocratic origins, as well as expanding the parameters of Evolutionary Feudal fiction.

Shiel opens his novel by acknowledging the established tropes of the aristocracy from which he is working—the literary and generic lineage set out by authors before him. The early villain of the text, Countess Clodagh, is several years older than her fiancé Adam Jeffson (himself a young, middle-class professional man); it is through Clodagh that we feel the weight, history, and cultural resonance of her class, a weight and history which will eventually be mirrored in Jeffson's one-man aristocracy later in the text. In speaking of their upcoming marriage, Jeffson says:

> Our proposed marriage was opposed by both my family and hers: by mine, because her father and grandfather had died in lunatic asylums; and by hers, because, forsooth, I was neither a rich nor a noble match. A sister of hers, much older than herself, had married a common country doctor, Peters of Taunton, and this so-called *mésalliance* made the so-called *mésalliance* with me doubly detestable in the eyes of her relatives. But Clodagh's extraordinary passion for me was to be stemmed neither by their threats nor prayers. What a flame, after all, was Clodagh! Sometimes she frightened me.[73]

Shiel introduces Clodagh as a plotting, malevolent aristocrat straight from annals of Gothic and radical fiction, who possesses an affinity for luxury and status reminiscent of the aristocrats in silver fork fiction, and with a family history that plays upon sensation fiction's fear of breeding, exogamy, and latent insanity. More significant is how widely the novel pathologises Clodagh, her family, and their history. Her family's mental health issues, as Jeffson illustrates, make her a dubious candidate for a bride and future mother. That Jeffson is willing to overlook the strong social opposition to their marriage—and to overlook it purely for the sake of her status—indicates a continuation of the contradictory discourses surrounding ideas of 'noble blood' and 'good breeding' that were brought up in the marriage debates and proto-eugenics rhetoric seen in our discussion of sensation fiction, as well as seen in other Evolutionary Feudal texts. Jeffson appears to be following a social pattern: Clodagh's father and grandfather both married and reproduced despite the public concern over their lineage and suitability, indicating that Darwinian breeding practices have become confused over time as they become embedded in other cultural practices. Unlike the aristocrats of *After London*, there is never any indication that her ancestors were of a 'natural' aristocracy. In fact, Clodagh is only referenced in terms of other aristocratic women, both historical and mythological, like Lucrezia Borgia, Medea, and Helen of Troy who are described by Jeffson in degenerative or dangerous terms, and yet were still socially desirable marriage partners through the luck of their birth.[74]

In addition to the potentially hereditary illness in Clodagh's family lies the further pathology of addiction. Clodagh's nephew Peter is a seasoned addict of the drug atropine, a circumstance which is treated disdainfully by Shiel as the result of an effete and over-cultured Aesthete with no concern for bodily survival or competition and, in fact, a flagrant disregard for the lucky conditions of his birth and body.[75] Peter does not appreciate his 'fine vitality, which so fitted him for an Arctic expedition', nor does he appreciate the seriousness of his medical condition, shrugging off doctors' opinions and regarding his injection of the drug as 'a mere flea-bite'.[76] Shiel goes so far as to write, '[i]t is he who has poisoned himself.'[77] Peter suffers from a timely overdose and nearly dies; his decision to continue using atropine after his recovery, despite his doctors' strict orders, is surprising to both Clodagh and reader alike and provides her with the opportunity to administer a second (this time fatal) overdose. It is this portrayed degeneracy of moral character and physical body alike which, as we will see, eventually leads to the destruction of the whole human race by providing an opportunity to the overly competitive Clodagh and Jeffson. This mass extinction is foreshadowed not only by Peter's destruction in which both he and Clodagh participate in the harm of their own line, but also Clodagh's physical form: Jeffson states that 'her reddish hair floated loose in a large flimsy cloud about her', and found that in addition to her causing him feelings of nausea, '[s]ometimes she frightened me'—he expresses the same feelings of fear and dizzying nausea later about the purple cloud.[78]

As I argue in my work on substance abuse as a Darwinian marker of class in this text,

> The most damning discourse is Clodagh's proximity to the medical and scientific community. She is heavily connected to, and yet continually at variance with, a doctor—her fiancé. A chemist in her own right, she only appears

to have the capacity for, or willingness to, make poisons: 'I find a sensuous pleasure, almost a sensual, in dabbling in delicate drugs'. Clodagh forwards the progression of scientific discovery, but her corrupt motives and interference lead to the destruction of thousands of years of human advancement and nearly all human life.[79]

Jeffson lukewarm reception to his fiancée solidifies his own position as a grasping social climber: despite his occasional glimmerings of terror, his feelings and actions towards Clodagh remain equally tepid both before and after she commits her murder. He scarcely spares her a thought on his voyage or in the face of global annihilation. In fact, the only benefit of their relationship seems to be her social status, something for which Jeffson early on expresses a particular weakness, as the owner of a medical practice which caters to the 'elite'.[80] Although a cold and perhaps misguided form of sexual selection, Darwin comments on the implicit connections between wealth, breeding, and evolution, saying that the accumulation of property and wealth provides an implicit advantage for the survival of one's children, and that 'without the accumulation of capital the arts could not progress; and it is chiefly through their power that the civilised races have extended, and are now everywhere extending their range, so as to take the place of the lower races'.[81] Darwin is, of course, only praising financial accumulation *in moderation*, enough to provide security and health. With Jeffson and Clodagh—much as was seen in *After London* and *The Time Machine*—wealth has mutated from a single consideration inside the practice of sexual selection to becoming the only consideration of a smaller and smaller pool. Elite groups become wealthier and tinier, leading to hereditary collapse; they no longer take the place of the 'lower races' but instead become them through, as Jeffson later calls it 'a true nobility [...] the nobility of self-extinction'.[82]

Despite Clodagh's dangerous lineage and destructive propensities, it is largely chance which leads Jeffson to becoming the world's sole and self-declared aristocrat. Clodagh's exploitation of the fortuitous circumstances surrounding her nephew places Jeffson in the position to reach the North Pole, but only with very long odds. On the voyage he snarls and snaps at the only other potential alpha-male on the trip, an electrician named David Wilson (who considers Jeffson to be 'a deadly beast') and it again through luck, rather than ability, that Jeffson is able to kill him in a duel.[83] Foreshadowing the randomness of the purple cloud, Wilson and Jeffson duel 'in clear, lavender light' and despite Wilson being 'much the best shot on the ship, and I an indifferent one', Wilson's shot somehow misses while Jeffson's hits its target. [84] All issues of luck aside, Jeffson is more adaptable and competitive than most of the crew, who seem to fail on every Darwinian level: the captain makes poor decisions regarding basic survival, which risk the entire expedition. A meteorologist named Aubrey Maitland engages in inappropriate buffoonery which gets him shot: he disguises himself in the pelt of a recently killed bear in an attempt to scare the crew as a joke. Maitland's inability to recognise danger—be it the danger that a bear presents to the human community, or the danger of a human community responding a perceived predator—leads to Jeffson accidentally shooting Maitland in what he considers to be defence of his own life and the life of the crew. Darwinian survival is acknowledged heavily in this section of the book, and Jeffson identified as an *After London*-esque 'natural' tribal-aristocrat.

Strangely, Jeffson's incredible good luck, which goes far beyond his considerable physical and mental abilities, is a recasting of the Carlylean aristocratic divinity. Jeffson's luck in the dual against Wilson and his survival at the Pole are never explained at all: scientists in the novel speculate that the purple cloud is rendered harmless in extremely cold temperatures; if Blavatsky is to be believed in the reality of the book, perhaps Jeffson stumbled on 'the existence of a fountain of life [...] in the North Pole' and did not recognise it.[85] In his introduction to *The Purple Cloud*, John Sutherland argues that 'Shiel is too canny a writer to make his novel a mere vehicle for Blavatsky's lunacies. But he draws on her ideas as a suggestive *décor*.'[86] Rather, Blavatsky's theories serve as juxtaposition to the large amounts of scientific data and meticulous and professional methodologies Jeffson expresses over the course of the novel, embodying the Darwinian and Carlylean debate over whether the origin of human roles is rooted in the physical, scientific world or in the spiritual or religious world. Jeffson himself does not fully understand his body's immunity to the toxin, and at the end of the novel he fears the cloud's return, convinced that he could not be lucky enough to survive exposure a second time.[87] Darwinian ideas of competition and strength are explored in the expedition to the Pole but are never fully brought to fruition, abandoned instead to focus on Darwin's notion of circumstance and environment, that 'we are apt to look at progress as the normal rule in human society; but history refutes this [...] Progress seems to depend on many concurrent favorable conditions.'[88] Jeffson and Leda's survival results largely from fortunate external circumstances. Indeed, their survival and future breeding potential leads to authorial, character, and audience confusion over the status of human progress and how this apocalyptic event will be portrayed in the annals of history: as a progression of the luckiest and fittest, as a devolution caused by a now-narrow genetic pool, or as a period of statis in which all of man's foibles and successes are repeated.

Shiel's notion of a 'luck of birth' or 'luck of body' aristocracy is most heavily reinforced by Jeffson's romantic partner in the second half of the novel, Leda. Although the only woman left on Earth, Leda leads a charmed life both before and after the apocalypse. As the daughter of a sultan, she is of a blood old and noble enough to appeal to Jeffson's snobbery, the same trait which paired him with the unappealing Clodagh. Indeed, trapped by his societal notions of blood, history, and aristocracy, Jeffson even calls Leda 'Clodagh' for more than a year before she illustrates to him her fundamental differences from his first fiancée and demands to be renamed. Clodagh was of an old aristocracy with an implied degeneracy and, while Leda is also of an old and even higher, dynastic line, Leda is also of a pure 'natural' aristocracy based solely upon survival, in the vein of *After London*'s earliest aristocrats. When she was an infant, Leda and her mother were sealed in a vault in her father's palace as the purple cloud descended, the air-tight vault protecting them from the poison of the cloud and leaving them with enough oxygen, food, and water inside to keep them alive long after the cloud had moved on. Leda's mother died, and Leda was left from a very young age to fend for herself with plenty of provisions remaining inside, and no predators available to harm her from the outside. While she would no doubt have died in a harsher environment, she possesses a lucky body that finds itself in the most beneficial circumstances at the most beneficial times. Jeffson says of her natural affinity for success, 'it is clear this creature is the *protégée* of someone, and favouritism is in the world.'[89] The walls of the vault eventually crumble and she is let out into the world where she immediately encounters Jeffson, who briefly considers killing and eating her before deciding instead to take her in and care for her.[90] Again, Leda's body finds itself in lucky circumstances, as she is bewildered by all life outside her vault and would likely have died had she not found a protector at that exact moment—and, indeed, had Jeffson not rather arbitrarily decided to turn from cannibalistic predator to adopted parent.

Leda not only embodies the randomness of an uncaring universe which, according to Shiel, places individuals at the top of the pinnacle through luck; she also embodies the fears and hopes caught up in *fin de siècle* investigation of human and class origins. Unlike the survivors in *After London* and in *The Time Machine* who are weighed down by the half-forgotten cultural baggage which leads to their evolution or devolution, Leda is entirely devoid of historical influences and palimpsests. She has grown up without culture and all that it entails, including language, material wealth, and relationships. Her life has been nothing but basic, individual survival, untainted by society or any other comparative marks of evolution. She is not, however, depicted as devolved or degenerate, but rather as pure: a form of life existing in nature, neither ascending nor descending the evolutionary hierarchy. Jeffson states that 'all she does [is] without effort: rather with the flighty naturalness with which a bird takes to the wing'.[91] While her luck and survival place her as the highest ranking (and only) woman on Earth, Leda demonstrates the potential to eradicate aristocracy through the eradication of culture.

She is without a social past and, until she meets Jeffson, her future mate, without a social future. As such, her body is presented as unpathologised until she spends more time with Jeffson, through whom she regains knowledge of the world and her own status within it. Indeed, the more cultured she becomes, the higher up the social ladder Jeffson perceives her to be and the more he medicalises her form. When he first spots her, Jeffson thinks, 'The earth was mine by old right: I felt that: and this creature a mere slave upon whom, without heat or haste, I might perform my will.'[92] Her body is largely without the consequences, influences, or coding of her lineage and she is merely a 'slave' to Jeffson's new aristocracy. But going hand-in-hand with their emerging relationship, and therefore the rebuilding of socio-cultural relations on Earth, is a continuation of the old concerns of class, lineage, and history that Jeffson came so close to eradicating. As their relationship progresses, Jeffson becomes increasingly obsessed with watching, culturing, and trying to understand Leda. Where once she was a mere slave beneath his contempt, or as natural as a bird taking to wing, now she is 'an object for the microscope', mimicking court ladies she has never seen, with 'airs and graces [which] are as natural to her as feathers to parrots [...] I could not have conceived of anyone in whom *taste* was a faculty so separate as in her, so positive and salient, like smelling or sight.'[93] What Jeffson now perceives to be natural to Leda are inherent manifestations of breeding, her class status coded on her body, and his need to investigate and examine her continuously. By this point Jeffson characterises his own body with 'monarch indignation [...] my neck stands stiff as sovereignty itself, and on my brow sits more than all the lordship of

Persepolis and Iraz', and yet he is the first rejuvenate proto-eugenicist stereotypes about the aristocratic body and breeding.[94] He says of Leda's romantic overtures:

> You are my Eve!—a little fool, a little piebald frog like you. But it will not do at all, at all! A nice race it would be with you for mother, and me for father, wouldn't it?—half-criminal like the father, half-idiot like the mother: just like the last, in short. They used to say, in fact, that the offspring of a brother and sister was always weak-headed: and from such a wedlock certainly came the human race, so no wonder it was what it was: and so it would have to be again now. Well no—unless we have the children, and cut their throats at birth: and *you* would not like that at all, I know, and, on the whole, it would not work.[95]

Although Jeffson's judgement is increasingly unreliable by this point in the narrative, his concerns about breeding are not necessarily arbitrary or ill-founded. The classical allusions, both to the Garden of Eden and to Leda and Zeus (Leda of *The Purple Cloud* being a queen in her own right, as the only surviving daughter of a sultan, and also the potential sexual partner to Jeffson, a self-proclaimed god), leave the reader feeling ambiguously about their potential union and offspring. It is unclear by the end of the novel if humanity will progress or regress through their children, whether they are going to rewrite a more positive version of the human race, or merely relive it. And, indeed, the main source of ambiguity about the future seems to lie in ambiguity about the past: man's error, which brought about the destruction of the human race, could either be scientific or moral, Darwinian or Carlylean.[96] Without exploring the past, one cannot learn from one's mistakes, and yet this delving into the unknown and unexamined seems to lead directly to man's downfall anyway. Jeffson, it seems, despite all of his classical and scientific knowledge, is just as muddled about human and class origin as anyone else. He says in the middle of his mania that he seemed 'to have gone right back to the very beginnings, and [resembled] man in his first, simple, gaudy conditions', but later in the text, after meeting Leda, states that he was 'greatly changed, God knows, from the portly inflated monarch-creature that strutted and groaned … so that my manner of life and thought might once more now have been called modern'.[97] His 'modern' style of life is both typical of the newer, restrained, middle-class mode of living in Britain characterised in *After London* as the antithesis of the aristocratic way of life, while also being characteristic of the simplicity, innocence, and class-lessness of prelapsarian life in the Garden of Eden, so heavily alluded to throughout the course of the novel. Jeffson is unable to characterise his own morality or class status accurately in terms of time, leading to muddled class and evolutionary identities frequently coupled with a sense of labyrinthine futility.

Jeffson's return to a less decadent way of living—be it pre-class or middle class, be it the most ancient or the most modern—brings the end of his madness and destructive tendencies, but it also leads to his dubious marriage, which may only produce further destruction. Jeffson states ambiguously—'Forward is no escape, nor backward, but *sideward* there may be a way!'—whatever that may mean for the logistics of his future with Leda.[98] Appropriately, the narrative ends without ever returning to its framing device of the seer in the trance. Just as we cannot locate Leda and Jeffson's children firmly on the scales of evolution, so we cannot be sure if their entire existence originates with the seer in the past, or if it is an ominous future reality.

Conclusion

Evolutionary Feudal texts foil Ruritanian ones by removing primogeniture from its divinely appointed pedestal and by denoting that class is not engendered or sustained through the grace of God, but rather through the functioning of nature and the development of culture. At the same time, the Evolutionary Feudal is both a companion to and reflection of Ruritania, in that the texts all satisfy or complicate the late nineteenth-century thirst for history, examine the role that an aristocrat's body serves for its subjects in reality and in literature, and imply that an aristocracy is a primal need of society, or at least is an institution that society cannot avoid. Further, all texts reveal a fatalism circling the topic of the aristocracy which is not exclusively the product of *fin de siècle* despondency. The culmination and synthesis of all bodily expectations throughout the Victorian era, especially with the onset of Darwinian thought, created a sense of impending finality surrounding all aristocratic mechanisms, a finality that overshadowed even the optimism and whimsy of Ruritania. Whether the aristocrats should be lauded or vilified was no longer the question; the question, more sharply than ever, became: would aristocrats survive?

And yet the outlook is perhaps not as pessimistic as the genre would initially have us believe. It was, after all, still a part of the Medieval Revival, which showcased endurance as much as decay. In fact, 'Decay and revival', argues Prys Morgan, 'are curiously intermixed, because very often those who bewail the decay were the very ones who brought about the revival'.[99] H.G. Wells, for example, is surprisingly more confident (however grimly so) about the longevity and evolutionary permutations of elite groups than Ruritania, *After London*, or any other text examined here. Yet, practically, some form of extinction or negative modification of the upper classes seemed inevitable, as evidenced by the aristocrats' continuingly self-imposed small population, even after the warnings of Darwin and others. Darwin writes, 'Rarity, as geology tells us, is the precursor to extinction [and] rare species will be less quickly modified or improved within any given period, and they will consequently be beaten in the race for life by the modified descendants of the commoner species.'[100] Undeniably, the aristocracy is a narrow class group and, with its history of political overthrow, defunct male lines, and the recreation or reinstatement of titles, it seemed in these texts to be trapped in a problematic, liminal state under the constant threat of extinction and the constant hope of rebirth.

With the onset of the 'future' at the end of the nineteenth century, which brought a feeling of death, impending peril or drastic change that Nordau described as the 'idea that the century is a kind of living being [...] passing through all the stages of existence [...] declining after blooming childhood, joyous youth, and vigorous maturity, to die with the expiration of the hundredth year, after being afflicted in its last decade with all the infirmities of mournful senility', it was only reasonable that late-Victorian society looked to the pseudo-medieval.[101] Working from the long traditions of Medievalism in the eighteenth and nineteenth centuries, the *fin de siècle* Medieval Revival could provide the comfort of nostalgia or could aid in the search for expectations of the future. In both instances, the aristocracy would naturally play an enormous role in the construction of either locus. For many who engaged with the medieval, especially in Ruritania's loose sense of the word, the existence of an aristocracy was one of the few certainties in a vague, fanciful or ignorant conception of the past.

Notes

1. For an overview of Victorian approaches to Darwinism, history, and the natural world, see Virginia Zimmerman's *Excavating Victorians* (Albany: State University of New York Press, 2008); John Batchelor's *H.G. Wells* (Cambridge: Cambridge University Press, 1995); Max Nordau's *Degeneration* (1892), 7th ed., translated from the 2nd ed. of the German work (New York: D. Appleton and Company, 1895); Bernard Bergonzi's

The Early H.G. Wells (Manchester: Manchester University Press, 1961); Steven McLean's *The Early Fiction of H.G. Wells* (Houndmills, Basingstoke: Palgrave Macmillan, 2009); Frank McConnell's *The Science Fiction of H.G. Wells* (Oxford: Oxford University Press, 1981); Justin Busch's *Utopian Vision*; Michael Draper's *H.G. Wells* (Houndmills, Basingstoke: Macmillan, 1987); Desmond Morris's *The Human Zoo* (London: Jonathan Cape LTD, 1969); Gillian Beer's *Darwin's Plots*; Michael Ruses's *Mystery of Mysteries*; and Joan DeJean's *Ancients Against Moderns* (Chicago: University of Chicago Press, 1997).

2. George Eliot, *Middlemarch* (1871) (Ware: Wordsworth Classics, 1994), p. 123.
3. See T.H. Huxley, "On the Method of Zadig" (1880) (*Collected Essays V. 4: Science and Hebrew Tradition.* Cambridge: CUP, 2011).
4. Evelyn Nesbit, *Prodigal Days* (1934), cited in Paula Uruburu's *American Eve: Evelyn Nesbit, Stanford White, the Birth of the "It" Girl, and the Crime of the Century* (New York: Riverhead Books, 2008), p. 1.
5. Zimmerman, p. 3.
6. Beer, p. 19.
7. McConnell, p. 59.
8. Thomas Carlyle, *On Heroes and Hero-Worship and the Heroic in History* (1841) 2 vols (London: Macmillan and Co., Limited, 1926), I, p. 15.
9. Q.S. Lilly, 'Primogeniture', *The Nineteenth-Century: A Monthly Review* 40.237 (Nov. 1896), pp. 765–68 (p. 768).
10. A.J.M. [Arthur J. Munby], 'Primogeniture', *Fraser's Magazine* 2.12 (December 1870), pp. 783–92 (p. 783).
11. For a few comprehensive overviews of Jefferies' work, see John Fowles's 'Introduction' to *After London* by Richard Jefferies (1885) (Oxford: Oxford University Press, 1980), pp. vii-xxi; W.J. Keith's *Richard Jefferies* (Toronto: University of Toronto Press, 1965), and Julian Wolfreys and William Baker's *Literary Theories* (Houndmills, Basingstoke: Macmillan, 1996).
12. Keith, p. 15.
13. David Blomfield, 'A Biography of Jefferies and a Note on the Manuscript', in *Literary Theories*, ed. by Julian Wolfreys and William Baker (Houndmills, Basingstoke: Macmillan, 1996), pp. 30–38 (p. 35); Fowles, p. ix.
14. Keith, p. 31.
15. Richard Jefferies, *After London, or Wild England* (London: Cassell & Company, Limited, 1885, repr. 1886), pp. 28–29.
16. Charles Darwin, *On the Origin of Species* (1859), ed. Jim Endersby (Cambridge: Cambridge University Press, 2009), p. 71.
17. Linda H. Peterson, 'Writing Nature at the Fin de Siècle: Grant Allen, Alice Maynell, and the Split Legacy of Gilbert White', *Victorian Review*, 36:2 (Fall 2010), pp. 80–91 (pp. 82–83).

18. Blomfield, p. 35; Peterson, p. 82.
19. Jefferies, *After London*, p. 60.
20. 'It is worth noting that Jefferies himself was continually emphasizing his own [political] impartiality. He may have been deceived in this belief, but it is clear enough that he did not look upon himself as a conscious propagandist' (Keith, p. 31).
21. Keith, p. 118.
22. Carlyle, *Heroes*, II, p. 15.
23. John Brannigan, 'A New Historicist Reading of "Snowed Up"', in *Literary Theories*, ed. by Julian Wolfreys and William Baker (Houndmills, Basingstoke: Macmillan, 1996), pp. 157–76 (p. 167).
24. Daniel P. Shea correlates Jefferies' leisurely plots as a manifestation of Jefferies' desire to slow down time and to keep his fictional England from progressing ('Richard Jefferies (1848–87)', *Victorian Review* 37:1 (Spring 20110), pp. 33–36, (pp. 34–35)). Though a discussion on Jefferies' views of technology is too large for the scope of this chapter, there is no question that Jefferies detested utilitarianism and the race for new technology (Blomfield 35; Fowles, p. vii; Jessica Maynard, 'Agriculture and Anarchy: A Marxist Reading of "Snowed Up"', in *Literary Theories*, ed. by Julian Wolfreys and William Baker [Houndmills, Basingstoke: Macmillan, 1996], pp. 129–56, p. 133). However, my analysis of *After London* shows that he held no such views about the progression of the natural world, of which man is a part.
25. Zimmerman, p. 15.
26. Jefferies, *After London*, p. 286.
27. Ibid., p. 125.
28. Ibid., p. 87.
29. Ibid., pp. 100–02.
30. Ibid., p. 57.
31. Anonymous, 'The Right of Primogeniture – Mr. Ewart's motion', *Tait's Edinburgh Magazine*, 8:4 (March 1837), pp. 159–62 (p. 162).
32. Charles Darwin, *The Descent of Man* (New York: D. Appleton and Company, 1871), p. 163.
33. Jefferies, *After London*, p. 126.
34. Ibid., p. 60.
35. Darwin, *Origin*, p. 82.
36. Jefferies, *After London*, p. 210.
37. Jefferies, *After London*, p. 206.
38. Ibid., p. 120.
39. Ibid., p. 91.
40. Darwin, *Descent*, p. 61.
41. Beer, p. 7.

42. McLean, p. 1; Adrian J. Desmond, 'Thomas Henry Huxley' in *Encyclopaedia Britannica*, Online Academic Edition (2014) http://www.britannica.com/EDchecked/topic/277746/Thomas-Henry-Huxley [Accessed 21 January 2014]; Patrick Parrinder, 'Wells, Herbert George (1866–1946)', in *Oxford Dictionary of National Biography*, ed. by H.C.G. Matthew and Brian Harrison (Oxford: Oxford University Press, 2004), Online ed., ed. by Lawrence Goldman (January 2011). https://doi.org/10.1093/ref:odnb/36831.
43. McConnell writes, 'Without Darwin there may literally not have been an "H.G. Wells" [...] evolutionary theory profoundly informed almost every aspect of his [Wells's] thought' (p. 53).
44. Batchelor, p. 1.
45. McConnell, pp. 18–19.
46. Ibid., p. 3.
47. Bergonzi, p. 50.
48. Batchelor, pp. 12–13; Anne-Julia Zwierlein, 'The Biology of Social Class', *Partial Answers: Journal of Literature and the History of Ideas,* 10:2 (June 2012), pp. 335–60 (p. 352); McLean, p. 27; Matthew Beaumont, 'Red Sphinx: Mechanics of the Uncanny in *The Time Machine*', *Science Fiction Studies*, 33:2 (July 2006), pp. 230–50 (p. 236).
49. Batchelor, p. 8.
50. H.G. Wells, *The Time Machine* (1895) ed. Patrick Parrinder (London: Penguin, 2005), p. 48.
51. Bergonzi, p. 48.
52. Martin Willis, *Mesmerists, Monsters, & Machines* (Kent, OH: The Kent State University Press, 2006), p. 206.
53. Wells, *Time Machine*, pp. 22–23.
54. Ibid., pp. 29–30.
55. James Eli Adams, *Dandies and Desert Saints* (Ithaca: Cornell University Press, 1995) p. 98.
56. Wells, *Time Machine*, pp. 57–58.
57. Ibid., p. 62.
58. Norbert Elias, *The Civilizing Process* (1939), trans. Edmund Jephcott (1978), ed. Eric Dunning, Johan Goudsblom, and Stephen Mennell (Oxford: Blackwell Publishers, 1994, repr. 2000), p. 100.
59. Wells, *Time Machine*, p. 62.
60. Ibid., p. 28.
61. Ibid., p. 30.
62. Ibid., p. 41.
63. Darwin, *Origin*, pp. 70–71; p. 76.
64. John Sutherland asserts that 'Shield drew heavily' on *After London*. John Sutherland, 'Introduction' to *The Purple Cloud* by M. P. Shiel (1901) (London: Penguin, 2012), pp. xiii–xxxiii (p. xxvi).

65. Monique R. Morgan states that it 'should come as no surprise that in this period of generic emergence and experimentation, Shiel produced a hybrid text that does not fit neatly into one specific category'. Monique R. Morgan, 'Madness, Unreliable Narration, and Genre in *The Purple Cloud*', *Science Fiction Studies* 36:2 (July 2009), pp. 266–83 (p. 266).
66. Sutherland, 'Introduction', p. xx.
67. Patrick Parrinder, 'From Mary Shelley to *The War of the Worlds*: The Thames Valley Catastrophe', *Anticipations: Essays on Early Science Fiction and its Precursors*, ed. by David Seed (Liverpool: Liverpool University Press, 1995) pp. 58–74 (p. 64).
68. M.P. Shiel, *The Purple Cloud* (1901), ed. John Sutherland (London: Penguin, 2012) p. 170; p. 147; p. 152.
69. For analysis of Shiel's interaction with and portrayals of race, see Sutherland's 'Introduction' to *The Purple Cloud*; William L. Svitavsky's 'From Decadence to Racial Antagonism: M.P. Shiel at the Turn of the Century', *Science Fiction Studies* 31:1 (Mar., 2004), pp. 1–24; Morgan's 'Madness, Unreliable Narration, and Genre in the Purple Cloud'; Brian W. Aldiss and David Wingrove, *Trillion Year Spree: The History of Science Fiction* (London: Victor Gollancz: 1986), pp. 145–146.
70. Ibid., p. 170.
71. Shiel, *The Purple Cloud*, p. 226.
72. Sutherland, 'Introduction', p. xxiv.
73. Shiel, *The Purple Cloud*, p. 16.
74. Ibid., p. 16; p. 20.
75. For a detailed reading on *The Purple Cloud*'s portrayal of substance abuse in relation to the Evolutionary Feudal, see my article '"Dabbling in Delicate Drugs": Aristocracy, Darwinism and Substance Abuse in M.P. Shiel's *The Purple Cloud*', *Victorians: A Journal of Culture and Literature* 142 (2022), pp. 63–84.
76. Shiel, *The Purple Cloud* p. 21; p. 22.
77. Ibid., p. 19.
78. Ibid., p. 129; p. 20; p. 16.
79. Boucher, 'Dabbling in Delicate Drugs', p. 77.
80. Shiel, *The Purple Cloud*, p. 12.
81. Darwin, *Descent*, p. 163.
82. Shiel, *The Purple Cloud*, p. 207.
83. Ibid., p. 33.
84. Ibid., p. 35; p. 34.
85. H. P. Blavatsky, *The Secret Doctrine: The Synthesis of Science, Religion, and Philosophy* (London: The Theosophical Publishing Company, 1888), II, p. 400.
86. Sutherland, 'Introduction', pp. xxiv–xxv.

87. Shiel, *The Purple Cloud*, pp. 258–60.
88. Darwin, *Descent*, p. 160.
89. Shiel, p. 217.
90. Ibid., pp. 187–89.
91. Ibid., p. 212.
92. Ibid., p. 187.
93. Ibid., p. 204; p. 238.
94. Ibid., p. 126.
95. Ibid., pp. 206–07.
96. Svitavsky, 'From Decadence to Racial Antagonism', p. 15.
97. Shiel, *The Purple Cloud*, p. 125; p. 239.
98. Ibid., p. 246.
99. Prys Morgan, 'From a Death to a View: The Hunt for the Welsh Past in the Romantic Period', *The Invention of Tradition* (1983), eds. Eric Hobsbawm and Terence Ranger (Cambridge: Cambridge University Press, rep. 2020), pp. 43–100 (p. 43).
100. Darwin, *Origin*, p. 92.
101. Nordau, p. 1.

Works Cited

Adams, James Eli, *Dandies and Desert Saints* (Ithaca: Cornell University Press, 1995).

A.J.M. [Arthur J. Munby], 'Primogeniture', *Fraser's Magazine* 2.12 (December 1870), pp. 783–92.

Aldiss, Brian W., and David Wingrove, *Trillion Year Spree: The History of Science Fiction* (London: Victor Gollancz: 1986), pp. 145–146.

Anonymous, 'The Right of Primogeniture – Mr. Ewart's motion', *Tait's Edinburgh Magazine,* 8:4 (March 1837), pp. 159–62.

Batchelor, John, *H.G. Wells* (Cambridge: Cambridge University Press, 1985).

Beer, Gillian, *Darwin's Plots* (London: Routledge & Kegan Paul, 1983).

Bergonzi, Bernard, *The Early H.G. Wells: A Study of the Scientific Romances* (Manchester: Manchester University Press, 1961).

Blavatsky, H.P., *The Secret Doctrine: The Synthesis of Science, Religion, and Philosophy* (London: The Theosophical Publishing Company, 1888).

Blomfield, David, 'A Biography of Jefferies and a Note on the Manuscript', in *Literary Theories*, ed. by Julian Wolfreys and William Baker (Houndmills, Basingstoke: Macmillan, 1996), pp. 30–38.

Boucher, Abigail, '"Dabbling in Delicate Drugs": Aristocracy, Darwinism and Substance Abuse in M. P. Shiel's *The Purple Cloud*', *Victorians: A Journal of Culture and Literature* 142 (2022), pp. 63–84.

Brannigan, John, 'A New Historicist Reading of "Snowed Up"', *Literary Theories*, ed. by Julian Wolfreys and William Baker (Houndmills, Basingstoke: Macmillan, 1996), pp. 157–76.

Busch, Justin E.A., *The Utopian Vision of H.G. Wells* (Jefferson, North Carolina: McFarland & Company, 2009).

Carlyle, Thomas, *On Heroes, Hero-Worship and the Heroic in History* (1841) 2 vols (London: Macmillan and Co., Limited, 1926).

Darwin, Charles, *The Descent of Man* (New York: D. Appleton and Company, 1871).

Darwin, Charles. *On the Origin of Species* (1859), ed. by Jim Endersby (Cambridge: Cambridge University Press, 2009).

DeJean, Joan, *Ancients Against Moderns: Culture Wars and the Making of a Fin de Siecle* (Chicago: University of Chicago Press, 1997).

Desmond, Adrian J., 'Thomas Henry Huxley', in *Encyclopaedia Britannica*, Online Academic Edition (2014), http://www.britannica.com/EDchecked/topic/277746/Thomas-Henry-Huxley [Accessed 21 January 2014].

Draper, Michael, *H.G. Wells* (Houndmills, Basingstoke: Macmillan, 1987).

Elias, Norbert, *The Civilizing Process* (1939), trans. Edmund Jephcott (1978), ed. Eric Dunning, Johan Goudsblom, and Stephen Mennell (Oxford: Blackwell Publishers, 1994, repr. 2000).

Eliot, George, *Middlemarch* (1871) (Ware: Wordsworth Classics, 1994).

Fowles, John 'Introduction', in *After London* by Richard Jefferies (1885) (Oxford: Oxford University Press, 1980), p. vii–xxi.

Huxley, T.H., 'On the Method of Zadig' (1880) (*Collected Essays* V. 4: *Science and Hebrew Tradition*. Cambridge: CUP, 2011).

Jefferies, Richard, *After London, or Wild England* (London: Cassell & Company, Limited, 1885, repr. 1886).

Keith, W.J., *Richard Jefferies* (Toronto: University of Toronto Press, 1965).

Lilly, Q.S., 'Primogeniture', *The Nineteenth-Century: A Monthly Review* 40.237 (Nov. 1896), pp. 765–68.

Maynard, Jessica, 'Agriculture and Anarchy: A Marxist Reading of "Snowed Up"', in *Literary Theories*, ed. by Julian Wolfreys and William Baker (Houndmills, Basingstoke: Macmillan, 1996), pp. 129–56.

McConnell, Frank, *The Science Fiction of H.G. Wells* (Oxford: Oxford University Press, 1981).

McLean, Steven, *The Early Fiction of H.G. Wells* (Houndmills, Basingstoke: Palgrave Macmillan, 2009).

Morgan, Monique R., 'Madness, Unreliable Narration, and Genre in *The Purple Cloud*', *Science Fiction Studies* 36:2 (July 2009), pp. 266–83.

Morgan, Prys, 'From a Death to a View: The Hunt for the Welsh Past in the Romantic Period', *The Invention of Tradition* (1983), eds. Eric Hobsbawm and Terence Ranger (Cambridge: Cambridge University Press, rep. 2020), pp. 43–100.

Morris, Desmond, *The Human Zoo* (London: Jonathan Cape LTD, 1969).

Nesbit, Evelyn, *Prodigal Days* (1934), cited in Paula Uruburu's *American Eve: Evelyn Nesbit, Stanford White, the Birth of the 'It' Girl, and the Crime of the Century* (New York: Riverhead Books, 2008).

Nordau, Max, *Degeneration* (1892), 7th ed., translated from the 2nd ed. of the German work (New York: D. Appleton and Company, 1895).

Parrinder, Patrick, 'From Mary Shelley to *The War of the Worlds*: The Thames Valley Catastrophe', *Anticipations: Essays on Early Science Fiction and its Precursors*, ed. by David Seed (Liverpool: Liverpool University Press, 1995) pp. 58–74.

Parrinder, Patrick.'Wells, Herbert George (1866–1946)' in *Oxford Dictionary of National Biography*, ed. by H.C.G. Matthew and Brian Harrison (Oxford: Oxford University Press, 2004), online ed. Ed. Lawrence Goldman (January 2011), https://doi.org/10.1093/ref:odnb/36831.

Peterson, Linda H., 'Writing Nature at the Fin de Siècle: Grant Allen, Alice Maynell, and the Split Legacy of Gilbert White', *Victorian Review*, 36:2 (Fall 2010), pp. 80–91.

Ruse, Michael, *Mystery of Mysteries: Is Evolution a Social Construction?* (Cambridge, MA: Harvard University Press, 1999).

Shea, Daniel P., 'Richard Jefferies (1848–87)', *Victorian Review*, 37:1 (Spring 2011), pp. 33–36.

Shiel, M.P., *The Purple Cloud* (1901), ed. John Sutherland (London: Penguin, 2012).

Sutherland, John, 'Introduction' to *The Purple Cloud* by M. P. Shiel (1901) (London: Penguin, 2012), pp. xiii–xxxiii.

Svitavsky, William L., 'From Decadence to Racial Antagonism: M.P. Shiel at the Turn of the Century', *Science Fiction Studies* 31:1 (Mar., 2004), pp. 1–24.

Wells, H.G., *The Time Machine* (1895), ed. Patrick Parrinder (London: Penguin, 2005).

Willis, Martin, *Mesmerists, Monsters & Machines* (Kent, OH: The Kent State University Press, 2006).

Wolfreys, Julian, and William Baker (eds.), *Literary Theories* (Houndmills, Basingstoke: Macmillan, 1996).

Zimmerman, Virginia, *Excavating Victorians* (Albany: State University of New York Press, 2008).

Zwierlein, Anne-Julia, 'The Biology of Social Class', *Partial Answers: Journal of Literature and the History of Ideas*, 10:2 (June 2012), pp. 335–60.

CHAPTER 7

Conclusion

In *The Court Society* (1969), Norbert Elias expounds at length about one of the major traps of studying class: that a discussion of systems often easily transforms into praise or censure of rulers.[1] Perhaps this is inevitable: aristocratic bodies, health, and lineage serve as highly visible textual objects which frequently operate in literature as an expression of cultural anxieties, desires, and expectations. More specifically, the literary figure of the aristocrat is a critically palimpsestic canvas upon which endless interpretations and readings may be cast and through which paradoxes may be untangled. Representations and interpretations of aristocracy not only serve to reveal what various class and social groups believe to be true of the 'elite'—despite Len Platt's argument that literature has 'always relished a *knowingly* falsifying account' of aristocracy without 'any real imperative toward "getting it right"' (whatever 'getting it right' means)—but also what these representations reveal other class and social groups believe to be true about themselves.[2]

Today, the aristocracy does not occupy quite as much cultural or political hegemony as they did in the nineteenth century. It is therefore tempting to think that this monograph might be the beginning and end of this realm of study—that, *if* this field ever needed to be opened up, the time to do so was long before now. But recent popular culture, genre fiction, and media engagement would indicate otherwise. Whatever the reality of

A. Boucher, *Science, Medicine, and Aristocratic Lineage in Victorian Popular Fiction*, Palgrave Studies in Literature, Science and Medicine, https://doi.org/10.1007/978-3-031-41141-0_7

aristocratic influence on the material world today, aristocratic bodies are still an enormously attractive space on which authors and directors can project contemporary preoccupations—especially pathological ones. In recent years, we've seen the adoration of TV shows like *Downton Abbey*, *Bridgerton*, and *Game of Thrones*.[3] There have been least three historical shows about the class-climbing Medici family in the last five years (*Reign*, *Medici*, and *The Serpent Queen*), which have dominated streaming services.[4] At the time of writing, there have been three major productions *in the last eighteen months alone* about Elisabeth of Austria, all of which received huge streaming numbers and/or critical acclaim. (*Corsage*, *The Empress*, and *Sisi*).[5] The dozens of best-selling historical novels by Philippa Gregory and Hilary Mantel (among others) find their historical plots mimicked in wildly popular modern incarnations of court intrigue, like *Succession*, *White Lotus*, and *Yellowstone*.[6] And this is to say nothing of the countless and ongoing adaptations of nineteenth-century novels which constantly renegotiate our understandings of aristocrats from a contemporary vantage. All of these texts deal overtly in pathologising the elite, in commenting on their health, (dis)ability, physical features, addictions, inherited traits, and their intersections with classed and raced bodies. Perhaps most pertinent of all, thought, has been the relentless coverage of the British royal family, discussions of which almost always turn to health, lineage, and (pseudo-) science. Their connections to various class groups, to endogamic or exogamic marriage, their weight and attractiveness, their pregnancies and parenting styles, their national and racial backgrounds, their assumed or diagnosed psychological states, their medical problems and lifespans—all have been picked apart, with little sign of stopping; as an institution they are still too pertinent a *literary* object to avoid being discussed.

In pure literary scholarship, there is still plenty of work to be done on aristocratic pathology and lineage. This monograph could not hope to make an exhaustive study of all the medical and scientific issues which appeared in fiction and resonated with large audiences, nor could it begin to cover all the popular literary genres that developed in the nineteenth century (let alone the genres that developed before, or after). Gothic, Romantic, drawing-room, detective, Weird, Decadent, Western, and children's fiction were scarcely touched on in this monograph, if even referenced at all. And this is to say nothing about more 'literary' texts in the realist or naturalist modes, or in poetry, or in plays, or in early cinema.

But if one may venture into a different realm of textual reading, we can add a richer dimension and greater applicability to the work done in this monograph. It's true that while aristocrats are still highly resonant figures in fiction, they are 'read' to a much smaller extent in the real world today, purely because the cultural power once held by the aristocrat has shifted to the celebrity. We see the same medicalised or pathologised language, the same policing of the body, and even the same focus on lineage (literal or professional or conceptual) applied to the contemporary celebrity as we did to the Victorian noble. This is not to simplify celebrity studies and star theory too much: there are many significant contextual, technological, economic, and sociological concerns which would caveat such a reading. But from a literary perspective, there *is* a great deal of overlap; there is more than a passing Wittgensteinian 'family resemblance' in how these two groups have developed and how they are read and consumed, especially in scientific and medical terms. Sociologist and journalist Francesco Alberoni locates contemporary stardom in 'kings and nobles [...] priests, prophets, and men of power', and that modern-day stars likewise form their own class, or at least their own distinct social group: a powerless elite 'whose institutional power is very limited or non-existent, but whose doings and way of life arouse a considerable and sometimes even a maximum degree of interest'.[7] The basic conditions for stardom to exist, Alberoni argues, are the very ones which saw the phasing out of aristocracy: a large-scale society in which social mobility is a possibility. Film scholar Richard Dyer, meanwhile, articulates that stardom comes from the combination of 'the spectacular with the everyday, the special with the ordinary'—much like aristocrats, especially in literature, stars exist to be cast as 'types' to fulfil certain roles for an audience.[8] It is the particular overlap of 'the special with the ordinary' implicit in the representation of aristocratic bodies that enabled audiences to use them as a litmus test for the medicalisation of all other bodies.

I have spent this monograph looking at the heredity—the metaphorical and literal lineage—of the aristocrat; it is here that we must consider their evolution into something new, and how gazing both backwards and forwards can open new dimensions for scholarship. The conditions and development of modern celebrity were already overlapping heavily with the aristocracy, at least as early as the silver fork novels, although likely far earlier (albeit in a different historical context). Elite groups have always had some form of curation of their image to enhance or define public ideals, assuage anxieties, or hide more provocative traits. With the rise of mass media, increased literacy rates, and various technologies like photography, film, and those that shortened travel, an aristocrat was able to be exposed to more and more of the general populace and to be read with fewer intermediaries, even as social and economic reforms reduced the material relevance of aristocrats to the general public. Celebrity was born in the aristocracy and evolved, perhaps somewhat bumpily, out of it; as celebrities are beholden to many of the aristocratic structures around fame, language, and image creation, so will much of the pathology associated with these structures trail after them. Today (and for much of the history of celebrity), one could scarcely open a popular magazine or watch an interview without being bombarded with medical and scientific issues about someone famous. The most common are, of course, are the supposed formulas and rituals undertaken to achieve a particular body type or beauty feature or to avoid ageing, but the medicalisation goes much deeper than this, as does the focus on lineage. Many stars, following exactly in the footsteps of nineteenth-century aristocrats, have become the faces of various 'wellness' trends or use their platforms to endorse products and procedures—a practice which also sustained both elite groups financially, increasing the lifespan of dwindling careers or reinvigorating a dying dynasty. Once married, celebrities are monitored for signs of pregnancy with the same fervour as any aristocrat potentially carrying a long-awaited heir, which many children of celebrities *are*: a high number of current celebrities (who are now called 'nepo' babies) were born into the system, grew up under a public lens, and made their debut much in the same way nineteenth-century aristocratic offspring. And, while under vastly improved conditions today, one does not have to go far back to see the careful management and hiding of various stigmatised 'pathologies', like queerness, mental health and other disabilities, or performances around race. In many star-making institutions, these practices were as deeply institutionalised as the rituals and shibboleths of the most exclusive royal court or social season. Tracing modern celebrity back to 'the king's two bodies' and understanding why we *implicitly* read them in terms of the body and health and pathology, will provide new resonances to both. By understanding the origins of celebrity codification, by conceptualising how and why public figures can manipulate their images, and how and why we are driven to pathologise them, can better help us predict what elite groups might develop going forward, and what this lineage might mean for genres and audiences in the future.

Notes

1. Norbert Elias, *The Court Society* (1969), trans Edmund Jephcott (Dublin: University College Dublin Press, 1983; repr. 2006), p. 26.
2. Len Platt, *Aristocracies of Fiction: The Idea of Aristocracy in Late-Nineteenth-Century and Early-Twentieth Century Literary Culture* (Westport, CT: Greenwood Press, 2001), p. xiv (italics mine).
3. Julian Fellows (Executive Producer), *Downton Abbey* [TV show], ITV (2010–2015); Julian Fellows (Executive Producer) and Michael Engler (Director), *Downton Abbey* [film], Carnival Films and Perfect World Pictures (2019); Julian Fellows (Executive Producer) and Simon Curtis (Director), *Downton Abbey: A New Era* [film], Universal Pictures and Focus Features (2022); Shonda Rhimes and Chris Van Dusen (Executive Producers), *Bridgerton* [TV show], Netflix (2020–present); David Benioff and D.B. Weiss (Executive Producers), *Game of Thrones* [TV show], HBO (2011–2019); George R.R. Martin and Ryan Condal (Executive Producers), *House of the Dragon*, HBO (2022–present).
4. Laurie McCarthy and Stephanie SenGupta [Executive Producers) *Reign* [TV show], Warner Bros. and CBS Television Studies (2013–2017; Frank Spotnitz (Executive Producer), *Medici* [TV show], Lux Vide and Rai Fiction (2015–2019); Stacie Passon and Justine Haythe (Executive Producers), *The Serpent Queen* [TV show], Starz (2022–present).
5. Marie Kreutzer (Director), *Corsage* [film] IFC Films and Picturehouse Entertainment, (2022); Jochen Laube and Fabian Maubach (Executive Producers), *The Empress* [TV show], Netflix (2022–present); Heinrich Ambrosch and Andreas Gutzeit (Executive Producers), *Sisi* [TV show], Story House Productions, Cinevilla Films, and Satel Film (2021–22).
6. Jesse Armstrong (Executive Producer), *Succession* [TV show], HBO (2018–present); Mike White (Executive Producer), *White Lotus* [TV show], HBO (2021–present); Taylor Sheridan and John Linson (Executive Producers), *Yellowstone* [TV show], Paramount Network (2018–present).
7. Francesco Alberoni, 'The Powerless "Elite": Theory and Sociological Research on the Phenomenon of the Stars', *Stardom and Celebrity: A Reader*, eds. Sean Redmond and Su Holmes (Thousand Oaks, CA: SAGE, 2017), pp. 65–78 (p. 65).
8. Richard Dyer, *Stars* (1979), 2nd ed. (London: BFI Publishing, 1998), p. 35.

Works Cited

Alberoni, Franceso, 'The Powerless "Elite": Theory and Sociological Research on the Phenomenon of the Stars', *Stardom and Celebrity: A Reader*, eds. Sean Redmond and Su Holmes (Thousand Oaks, CA: Sage, 2017), pp. 65–78.

Ambrosch, Heinrich and Andreas Gutzeit (Executive Producers), *Sisi* [TV show], Story House Productions, Cinevilla Films, and Satel Film (2021–22).

Armstrong, Jesse (Executive Producer), *Succession* [TV show], HBO (2018–present).

Benioff, David and D.B. Weiss (Executive Producers), *Game of Thrones* [TV show], HBO (2011–2019).

Dyer, Richard, *Stars* (1979), 2nd ed. (London: BFI Publishing, 1998).

Elias, Norbert, *The Court Society* (1969), trans. Edmund Jephcott (Dublin: University College Dublin Press, 1983; repr. 2006).

Fellows, Julian (Executive Producer), *Downton Abbey* [TV show], ITV (2010–2015).

Fellows, Julian (Executive Producer) and Michael Engler (Director), *Downton Abbey* [film], Carnival Films and Perfect World Pictures (2019).

Fellows, Julian, (Executive Producer) and Simon Curtis (Director), *Downton Abbey: A New Era* [film], Universal Pictures and Focus Features (2022).

Kreutzer, Marie (Director), *Corsage* [film] IFC Films and Picturehouse Entertainment, (2022).

Laube, Jochen and Fabian Maubach (Executive Producers), *The Empress* [TV show], Netflix (2022–present).

Martin, George R.R. and Ryan Condal (Executive Producers), *House of the Dragon*, HBO (2022–present).

McCarthy, Laurie and Stephanie SenGupta [Executive Producers) *Reign* [TV show], Warner Bros. and CBS Television Studies (2013–2017).

Passon, Stacie, and Justine Haythe (Executive Producers), *The Serpent Queen* [TV show], Starz (2022–present).

Platt, Len, *Aristocracies of Fiction: The Idea of Aristocracy in Late-Nineteenth-Century and Early-Twentieth Century Literary Culture* (Westport, CT: Greenwood Publishing, 2001).

Rhimes, Shonda and Chris Van Dusen (Executive Producers), *Bridgerton* [TV show], Netflix (2020–present).

Sheridan, Taylor and John Linson (Executive Producers), *Yellowstone* [TV show], Paramount Network (2018–present).

Spotnitz, Frank (Executive Producer), *Medici* [TV show], Lux Vide and Rai Fiction (2015–2019).

White, Mike (Executive Producer), *White Lotus* [TV show], HBO (2021–present).

Index[1]

A
Anatomy Act (1832), 78
Apothecaries Act (1815), 47
Aristocracy, definition of, 193
See also Nobility, definition of; Upper class, definition of
Aspiration, 8, 46, 49

B
Beddoes, Thomas, 37
Beer, Gillian, 7, 159, 189, 200
Berger, John, 8
Body politic, 5, 7, 36, 93, 170
Boswell, James, 37
Bourdieu, Pierre, 4, 32, 40
See also Habitus
Brontë, Charlotte, 113, 117
Brontë, Emily, 113
Byron, Lord (George Gordon), 38

C
Cannadine, David, 3, 12, 38
Carlyle, Thomas, 34, 157–160, 163, 167, 169, 172, 173, 175, 179, 181n14, 182n41, 190, 193, 194, 197
On Heroes and Hero-Worship (1841), 158, 181n14, 190
Sartor Resartus (1836), 34, 157, 158
Celebrity, 4, 6, 8, 32, 43, 44, 48, 229, 230
Chambers, Robert, 159
Chartism, 79, 80
Cheyne, George, 37
Consumption, 28, 31, 38, 41, 43, 46, 54, 56, 205
Corn Laws, 39
Cousin-marriage, 125, 129, 141, 142, 144
See also Endogamy; Exogamy; Incest

[1] Note: Page numbers followed by 'n' refer to notes.

A. Boucher, *Science, Medicine, and Aristocratic Lineage in Victorian Popular Fiction*, Palgrave Studies in Literature, Science and Medicine, https://doi.org/10.1007/978-3-031-41141-0

D
Darwin, Charles, 115, 124, 125, 129, 145, 146n15, 155, 157–160, 164, 167, 169, 175, 181n21, 188, 192, 197–199, 201, 212, 213, 218, 221n43
  The Descent of Man (1871), 115, 159
  natural selection, 125, 159, 192, 197
  On the Origin of Species (1859), 115, 146n15, 159
  sexual selection, 125, 159, 160, 199, 205, 212
  survival of the fittest, 159, 160, 181n21, 196
Darwin, Erasmus, 83, 90, 159
Degeneration, 156, 170, 171, 173, 177, 178, 191, 194, 202
Dickens, Charles, 45, 113, 117
  Bleak House (1853), 113, 117
  A Tale of Two Cities (1859), 45
Doctor
  clinician, 134
  physician, 37, 44, 46, 50, 54, 56, 57, 92
  quack, 19, 45, 46, 56

E
Elias, Norbert, 5, 156, 157, 161, 205, 227
  The Civilizing Process (1939), 5, 156, 205
  The Court Society (1969), 5, 227
Endogamy, 19, 112, 116, 119–144, 198
  See also Cousin-marriage; Exogamy; Incest
Essentialism, 11, 12, 83
Eugenics, 10, 19, 113, 145n5
Evolution, 20, 148n38, 155–179
  genetics, 166
  hard hereditarianism, 162
  soft hereditarianism, 126, 162
  See also Chambers, Robert; Darwin, Charles; Darwin, Erasmus; Huxley, T.H.; Lamarck, J.B.; Lyell, Charles; Russel Wallace, Alfred
Evolutionary Feudal
  Jefferies, Richard, 188, 191–194, 196–199, 202, 206, 207, 220n20, 220n24
  Shiel, M.P., 188, 207, 208, 210, 211, 213, 214, 222n65
  Wells, H.G., 188, 189, 194, 200–203, 205–207, 218, 221n43
Exogamy, 19, 119, 120, 123, 126, 128, 138, 143, 144, 210
  See also Cousin-marriage; Endogamy; Incest

F
Family resemblances, 5, 15, 117, 118, 163, 229
  See also Wittgenstein, Ludwig
Fertility
  emasculation, 86, 87, 91–101
  feminisation, 86–91, 93–101
  impotence, 82, 90, 91, 94
  infertility, 77, 79, 82, 83, 90, 91, 93, 95
Fin de siècle, 5, 16, 17, 131, 155, 156, 158, 160, 167, 170, 171, 173, 178, 188, 189, 201, 214, 217, 218
Foucault, Michel, 7, 8, 40
  The Birth of the Clinic (1963), 7, 8
  Discipline and Punish (1975), 8
  History of Sexuality (1976-2018), 40
Frazer, James, 5, 174, 179
  See also The Golden Bough

G
Galton, Francis, 125–127, 129, 140, 148n42
Gaze, 4, 5, 7–12, 134, 149n59, 171
 See also Visibility
Genre fiction, definition of, 14
 See also Popular fiction, definition of
Gentlemen, concept of, 64
George IV, 39, 79, 86, 104n18
 See also Prince Regent
The Golden Bough, 5
 See also Frazer, James
Gothic, 2, 17, 78, 79, 117, 210, 228
Great Reform Act of 1832, 39
Greenwood, James, 79

H
Habitus, 4, 32, 39, 40, 43, 49, 84, 116, 134
 See also Bourdieu, Pierre
Healthcare industry
 patented medicines, 46
 quacks, 19, 45, 46, 56
Heredity
 genetics, 19, 113, 123, 124, 126, 128, 129, 131, 133, 144, 145, 167, 168, 177, 202, 214
 hard hereditarianism, 127–130, 133, 168
 soft hereditarianism, 125–128
Hobsbawm, Eric, 6, 111, 122
 See also Invented tradition
Homosexuality, 90, 105n34, 203
 See also Queerness
Huth, Alfred Henry, 136, 149n67
Huxley, T.H., 188, 201

I
Illness
 as fashion, 44
 as plot device, 19, 35, 41–43
 Sontag, Susan, 28, 35
Incest, 114, 119–121, 123, 136
 See also Cousin-marriage; Endogamy; Exogamy
Invented tradition, 6, 111, 112, 122
 See also Hobsbawm, Eric

K
Kantorowicz, Ernst H., 5
 See also King's Two Bodies
King's Two Bodies, 5
 See also Kantorowicz, Ernst H.

L
Lamarck, J.B., 124, 125, 159
Lancet, The, 47, 102, 140
Lower class, 12, 13, 19, 43, 51, 65, 66, 80, 87, 94, 99, 103n7, 120, 132, 133, 135, 136, 163, 172, 189, 191, 192
 See also Working class
Lyell, Charles, 159

M
Mansel, Henry Longueville, 114, 115, 117
Marriage Act (1835), 121
Marriage market, 35, 41, 43, 56, 121, 125
Masculinity, 60, 81–87, 94–96, 142, 170, 176–178, 195, 196
Mayhew, Henry, 80
McLennan, John Ferguson, 119, 120, 136, 139
 See also Primitive Marriage
Mearns, Reverend Andrew, 136, 139
Medical Registration Act (1858), 47
Medicine, definition of, 10
Medieval Revival, 156, 169, 187, 189, 217, 218
Melville, Herman, 113, 120
Mendel, Gregor, 129

Middle class, 1, 2, 4, 6, 8, 12, 19, 30–32, 37–41, 44–46, 51–54, 57, 60, 61, 63–65, 78, 81, 85, 89, 90, 95, 102, 118, 122–124, 130, 131, 134, 136, 139, 161–164, 166, 172, 182n41, 191, 194, 198–203, 210, 216, 217
Montaigne, Michel de, 37

N
Nesbit, Evelyn, 188
Nightingale, Florence, 123
Nobility, definition of, 3
See also Aristocracy, definition of; Upper class, definition of
Nordau, Max, 155, 218

P
Pathology, definition of, 11
Penny dreadful, 78
See also Penny fiction
Penny fiction, 17, 18, 77–79, 101, 112
See also Penny dreadful
Percival, Thomas, 47
Pharmacy Act of 1868, 47
Phrenology, 10
Physiognomy, 10, 155–179
Popular fiction, definition of, 13
See also, Genre fiction, definition of
Post-apocalyptic, 18, 20, 187, 200, 207
Primitive Marriage, 119
See also McLennan, John Ferguson
Primogeniture, 19, 35, 41, 43, 54, 55, 77, 79, 82, 83, 163, 176, 190, 193, 196–199, 202, 203, 217
Prince Albert, 115
Prince Regent, 79, 86, 91, 94–98, 100, 101
See also George IV
Protestant Reformation, 121, 122
Protestant work ethic, 85, 95, 166
See also Smiles, Samuel; Weber, Max
Psychology, 11, 19, 115

Q
Quackery, 45, 47, 48
Queerness, 84, 230
See also Homosexuality

R
Race, 2, 97, 112, 113, 124, 126, 134–144, 145n5, 149n76, 164, 166, 172, 177, 194, 200–202, 207, 208, 211, 212, 216, 218, 220n24, 222n69, 230
Radical fiction, 210
Reynolds, G.W.M., 18, 19, 77–102, 112, 162, 163, 170, 190
The Mysteries of London, 78
The Mysteries of the Court of London, 19, 77–102
Roman à clef, 32, 34
Rowlandson, Thomas, 38, 45
Ruritania
Hodgson Burnett, Frances, 164, 174
Hope, Anthony, 160, 161, 164–169, 173
McCutcheon, G.B., 181n22
Stevenson, Robert Louis, 164, 169–173
Ruskin, John, 85, 114
Russel Wallace, Alfred, 159

S
Science, definition of, 9
Sensation fiction
Alcott, Louisa May, 134
Braddon, Mary Elizabeth, 116, 117, 127, 132–134

Collins, Wilkie, 116–118, 120, 127, 128, 130–132, 138–142
Conan Doyle, Sir Arthur, 128
Eliot, George, 113, 137
Le Fanu, Sheridan, 143
Ouida, 126
'Rita' [Eliza Humphreys], 146n14
Wood, Mrs Henry [Ellen], 116, 117, 122, 130, 131, 134, 149n59
Sensibility, 19, 35–44, 48, 55, 56, 60, 62, 63, 111, 112
Silver fork fiction, 77, 80, 112, 210
Blair, Mrs Alexander ('Lady Humdrum'), 46
Blessington, Lady, 33, 40
Bulwer-Lytton, Edward, 6, 19, 29, 30, 32, 33, 46, 48–60, 62
Bulwer Lytton, Rosina, 19, 29, 32, 46, 48, 57–62
Burdett, C.D., 40
Bury, Lady Charlotte, 33, 46, 49
Colburn, Henry, 30, 46, 47
Disraeli, Benjamin, 33, 46, 50, 201
Gore, Catherine, 19, 29, 33, 40, 41, 48, 62–65, 72n103
Hazlitt, William, 34
Hook, Theodore, 33, 46, 50
Lamb, Lady Caroline, 46
Landon, Letitia, 42, 46, 47
Lewis, Lady Theresa, 4, 46
Lister, Thomas Henry, 31, 33, 46, 50
Morgan, Lady Sydney, 46
Normanby, Lord, 46
Pigott, Harriet, 46
Plumer Ward, Robert, 33, 46, 50
publishing industry, 36, 44
Sullivan, Arabella Jane, 46
Thackeray, William Makepeace, 27, 28
Trimmer Moore, Charlotte, 39, 42, 48
Smiles, Samuel, 95
See also Protestant work ethic
Social season, 112, 122, 230
Sociology, 19, 191
Spencer, Herbert, 126, 127, 159, 181n21

T

Trollope, Anthony, 113

U

Upper class, definition of, 202
See also Aristocracy, definition of; Nobility, definition of

V

Victoria, Queen, 39, 207
Visibility, 4, 8
See also Gaze

W

Wakley, Thomas, 47
Weber, Max, 4, 95
Weismann, August, 126, 127, 168
William IV, 39
Wittgenstein, Ludwig, 15, 117
See also Family resemblances
Working class, 1, 2, 13, 37, 77–81, 85, 88, 115, 162, 205
See also Lower class

Printed by Printforce, the Netherlands